بسم الله الرحمن الرحيم

الله
رسول
محمد

ABOUT THE AUTHOR

Under the pen-name HARUN YAHYA, the author has published many books on political and faith-related issues. An important body of his work deals with the materialistic world view and the impact of it in world history and politics. (The pen-name is formed from the names 'Harun' [Aaron] and 'Yahya' [John] in the esteemed memory of the two Prophets who struggled against infidelity.)

His works include The 'Secret Hand' in Bosnia, The Holocaust Hoax, Behind the Scenes of Terrorism, Israel's Kurdish Card, A National Strategy for Turkey, Solution: The Morals of the Qur'an, Darwin's Antagonism Against the Turks, The Evolution Deceit, Perished Nations, The Golden Age, Allah's Artistry in Colour, Glory is Everywhere, The Truth of the Life of This World, Confessions of Evolutionists, The Blunders of Evolutionists, The Dark Magic of Darwinism, The Religion of Darwinism, The Qur'an Leads the Way to Science, The Real Origin of Life, Miracles of the Qur'an, The Design in Nature, Self-Sacrifice and Intelligent Behaviour Models in Animals, Eternity Has Already Begun, Children Darwin Was Lying!, The End of Darwinism, Deep Thinking, Timelessness and the Reality of Fate, Never Plead Ignorance, The Miracle of the Atom, The Miracle in the Cell, The Miracle of the Immune System, The Miracle in the Eye, The Creation Miracle in Plants, The Miracle in the Spider, The Miracle in the Ant, The Miracle in the Gnat, The Miracle in the Honeybee.

Among his booklets are The Mystery of the Atom, The Collapse of the Theory of Evolution: The Fact of Creation, The Collapse of Materialism, The End of Materialism, The Blunders of Evolutionists 1, The Blunders of Evolutionists 2, The Microbiological Collapse of Evolution, The Fact of Creation, The Collapse of the Theory of Evolution in 20 Questions, The Biggest Deception in the History of Biology: Darwinism.

The author's other works on Quranic topics include: Ever Thought About the Truth?, Devoted to Allah, Abandoning the Society of Ignorance, Paradise, The Theory of Evolution, The Moral Values of the Qur'an, Knowledge of the Qur'an, Qur'an Index, Emigrating for the Cause of Allah, The Character of Hypocrites in the Qur'an, The Secrets of the Hypocrite, The Names of Allah, Communicating the Message and Disputing in the Qur'an, The Basic Concepts in the Qur'an, Answers from the Qur'an, Death Resurrection Hell, The Struggle of the Messengers, The Avowed Enemy of Man: Satan, Idolatry, The Religion of the Ignorant, The Arrogance of Satan, Prayer in the Qur'an, The Importance of Conscience in the Qur'an, The Day of Resurrection, Never Forget, Disregarded Judgements of the Qur'an, Human Characters in the Society of Ignorance, The Importance of Patience in the Qur'an, General Information from the Qur'an, Quick Grasp of Faith 1-2-3, The Crude Reasoning of Disbelief, The Mature Faith, Before You Regret, Our Messengers Say, The Mercy of Believers, The Fear of Allah, The Nightmare of Disbelief, Prophet Isa Will Come, Beauties Presented by the Qur'an for Life, The Iniquity Called "Mockery", The Secret of the Test, The True Wisdom According to the Qur'an, The Struggle with the Religion of Irreligion, The School of Yusuf, The Alliance of the Good, Slanders Against Muslims Throughout History, The Importance of Following Good Word, Why Do You Deceive Yourself?, Bouquet of the Beauties of Allah 1-2-3-4.

THE CREATION OF THE UNIVERSE

HARUN YAHYA

AL-ATTIQUE

Title: The Creation of the Universe
Author: Harun Yahya
Translated By: Robert Bragner
Copyright: All rights reserved
Printed: 2000
Printing supervised by: M.R.Attique
Printed at : Toronto – Canada

First Published by Vural Yayıncılık, İstanbul, Turkey in August 1999

{C} Al-Attique Publishers Inc.Canada 2000
ISBN : 1-894264-38-X

Published by: Al-Attique Publishers Inc.Canada
65-Treverton Drive Tel: (416) 615-1222
Scarborough Ont. Fax: (416) 615-0375
M1K 3S5 CANADA
E-mail: quran@istar.ca Website: www.al-attique.com
E-mail: al-attique@al-attique.com

REPRESENTATIVE IN USA
Islamic Education & Media
730 East 10th street, C.F,
Brooklyn, NY 11230
T+F: (718) 421-5428

DISTRIBUTOR IN SAUDI ARABIA
Dar-Al-Hadyan Publishers & Distributors
P/O Box No : 15031
Al-Riyadh:11444
T+F (966) 1-463-1685

Branch in Pakistan:
89 Qamer st People Colony
Shahdara Lahore
T+F : 9242-791-1678

Website: www.harunyahya.org - www.harunyahya.com
www.harunyahya.net email: info@harunyahya.org

AL-ATTIQUE

TO THE READER

● The reason why a special chapter is assigned to the collapse of the theory of evolution is that this theory constitutes the basis of all anti-spiritual philosophies. Since Darwinism rejects the fact of creation, and therefore the existence of Allah, during the last 140 years it has caused many people to abandon their faith or fall into doubt. Therefore, showing that this theory is a deception is a very important duty, which is strongly related to the religion. It is imperative that this important service be rendered to everyone. Some of our readers may find the chance to read only one of our books. Therefore, we think it appropriate to spare a chapter for a summary of this subject.

● Another point to be stressed is related to the content of the book. In all the books by the author, faith-related issues are told in the light of the Qur'anic verses and people are invited to learn Allah's words and to live by them. All the subjects that concern Allah's verses are explained in such a way as to leave no room for doubt or question marks in the reader's mind. The sincere, plain and fluent style employed ensures that everyone of every age and from every social group can easily understand the books. This effective and lucid way of recounting makes the books read suitable for reading in a single sitting. Even those who rigorously reject spirituality are influenced by the facts recounted in these books and cannot refute the truthfulness of their contents.

● This book and all the other works of the author can be read by individuals or studied in a group at a time of conversation. The reading of the books by a group of readers willing to profit from them will be useful in the sense that readers can relate their own reflections and experiences to one another.

● In addition, it will be a great service to the religion to contribute to the presentation and reading of these books, which are written solely for the good pleasure of Allah. All the books of the author are extremely convincing. For this reason, for those who want to communicate the religion to other people, one of the most effective methods is to encourage them to read these books.

● There are important reasons why reviews of other books by the author are added at the end of this book. Seeing them, the reader who takes this book in his hand will discover that there are many other books that share the same qualities as this book, and which we hope would also be a pleasure for him to read. He will see that there is a rich source of material on faith-related issues, of which he can make use.

● In these books, you will not find, as in some other books, the personal views of the author, explanations based on dubious sources, styles that are unobservant of the respect and reverence due to sacred subjects, nor hopeless, doubt-creating, and pessimistic accounts that create deviations in the heart.

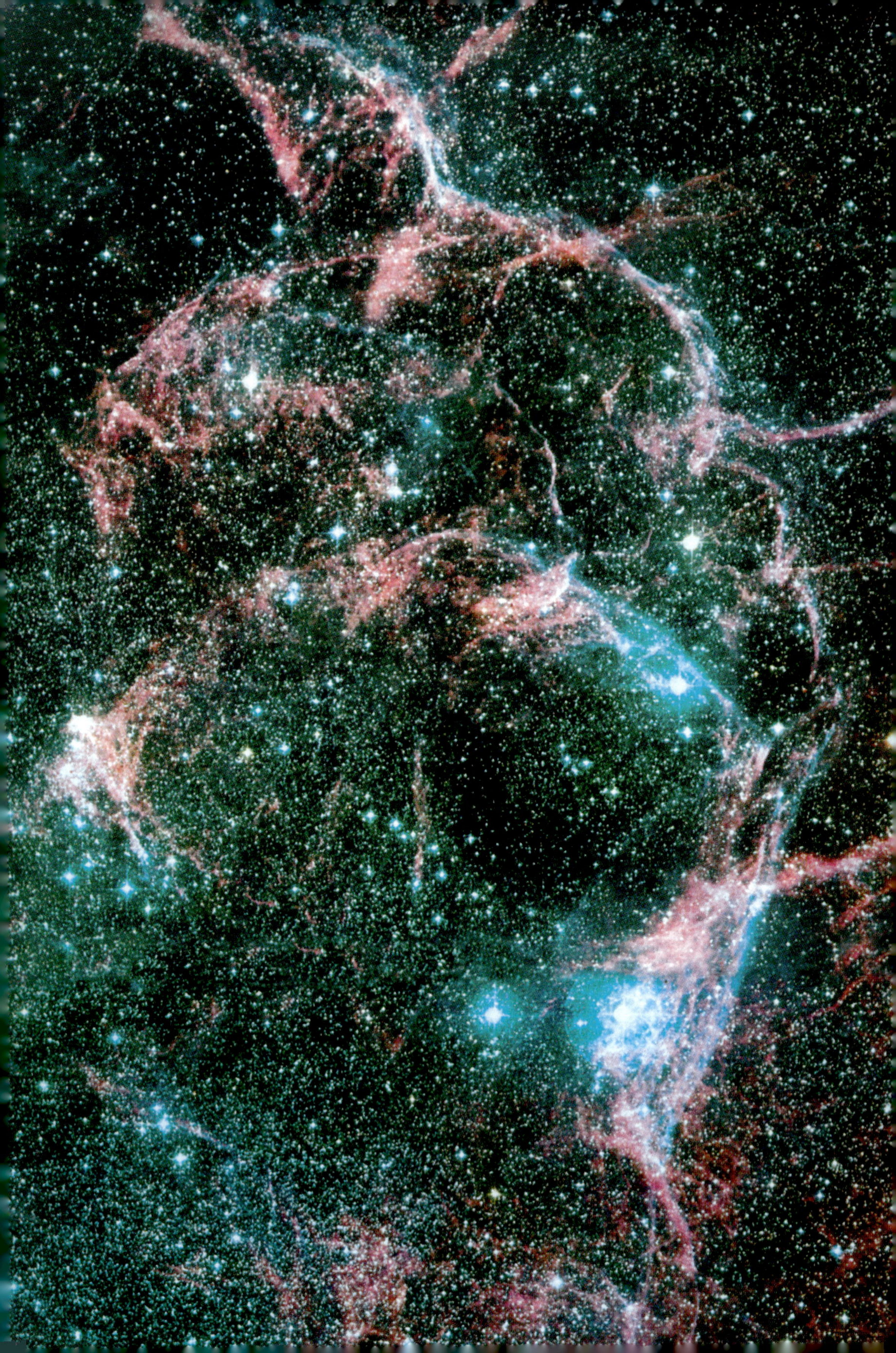

TABLE OF CONTENTS

INTRODUCTION

THE SCIENTIFIC COLLAPSE OF MATERIALISM

Materialism can no longer claim to be a scientific philosophy.

Arthur Koestler, the renowned Social Philosopher[1]

How did the endless universe we live in come into being? How did the equilibrium, harmony, and order of this universe develop?

How is it that this Earth is such a fit and sheltering place for us to live in?

Questions such as these have attracted attention since the dawn of the human race. The conclusion reached by scientists and philosophers searching for answers with their intellects and common sense is that the design and order of this universe are evidence of the existence of a supreme Creator ruling over the whole universe.

This is an indisputable truth that we may reach by using our intelligence. Allah declares this reality in His holy book, the Qur'an, which He inspired as a guide for humanity fourteen centuries ago. He states that He has created the universe when it was not, for a particular purpose, and with all its systems and balances specifically designed for human life.

Allah invites people to consider this truth in the following verse:

Are you stronger in structure or is heaven? He built it. He raised its vault high and made it level. He darkened its night and brought forth its morning light. After that He smoothed out the earth… (Surat an Naziat: 27-30)

Elsewhere it is declared in the Qur'an that a person should see and consider all the systems and balances in the universe that have been created for him by Allah and derive a lesson from his observations:

He has made night and day subservient to you, and the sun and moon and stars, all subject to His command. There is certainly Signs in that for people who pay heed. (Surat an-Nahl: 12)

In yet another verse of the Qur'an, it is pointed out:

He makes night merge into day and day merge into night, and He has made the sun and moon subservient, each one running until a specified time. That is Allah, your Lord. The Kingdom is His. Those you call on besides Him have no power over even the smallest speck.(Surah Fatir: 13)

This plain truth declared by the Qur'an is also confirmed by a number of the important founders of the modern science of astronomy. Galileo, Kepler, and Newton all recognised that the structure of universe, the design of the solar system, the laws of physics and their states of equilibrium were all created by Allah and they arrived at that conclusion as a result of their own research and observations.

Materialism: A 19th-Century Fallacy

The reality of the creation of which we speak has been ignored or denied since the earliest times by a particular philosophical point of view. It is called "materialism". This philosophy, which was originally formulated among the ancient Greeks, has also made an appearance from time to time in other cultures and has been advanced by individuals as well. It holds that matter alone exists and that it has done so for an infinity of time. From these tenets, it claims that the universe has also "always" existed and was not created.

In addition to their claim that the universe exists in an infinity of time, materialists also assert that there is no purpose or aim in the universe. They claim that all the equilibrium, harmony and order that we see around us are merely the product of coincidence. This "coincidence assertion" is also put forward when the question of how human beings came into being comes up. The theory of evolution, widely referred to as Darwinism, is another application of materialism to the natural world.

We just mentioned that some of the founders of modern science were faithful people who were in agreement that the universe was created and organised by Allah. In the 19th century, an important change took place in the attitudes of the scientific world with respect to this matter. Materialism was deliberately introduced to the agenda of modern science by various groups. Because the 19th century's political and social conditions formed a good basis for materialism, the philosophy gained wide acceptance and spread throughout the scientific world.

The findings of modern science however undeniably demonstrate how false the claims of materialism really are.

The Findings of 20th-Century Science

Let us recall the two assertions of materialism about the universe:

The universe exists in infinite time and, because it has no beginning or end, it was not created.

Everything in this universe is merely the result of chance and not the product of any intentional design, plan, or vision.

Those two notions were boldly advanced and ardently defended by 19th-century materialists, who of course had no recourse other than to depend upon the limited and unsophisticated scientific knowledge of their day. Both have been utterly refuted by the discoveries of 20th-century science.

The first to be laid in the grave was the notion of the universe existing in infinite time. Since the 1920s, there has been mounting evidence this cannot be true. Scientists are now certain that the universe came into being from nothingness as the result of an unimaginably huge explosion, known as the **"Big Bang"**. In other words, the universe came into being–or rather, it was created by Allah.

Modern science proves the reality of the creation of the universe by Allah, contrary to what outdated materialist philosophy maintains. Newsweek made "Science Finds God" the cover story of its July 27th 1998 issue.

The 20th century has also witnessed the demolition of the second claim of materialism: that everything in the universe is the result of chance and not design. Research conducted since the 1960s consistently demonstrates that all the physical equilibriums of the universe in general and of our world in particularly are intricately designed to make life possible. As this research deepened, it was discovered each and every one of the laws of physics, chemistry, and biology, of the fundamental forces such as gravity and electromagnetism, and of the details of the structure of atoms and the elements of the universe has been precisely tailored so that human beings may live. Scientists today call this extraordinary design the "anthropic principle". This is the principle that every detail in the universe has been carefully arranged to make human life possible.

To sum up, the philosophy called materialism has been utterly refuted by modern science. From its position as the dominant scientific view of the 19th century, materialism collapsed into fiction in the 20th.

How could it have been otherwise? As Allah indicates **"We did not create heaven and earth and everything between them to no purpose. That is the opinion of those who are disbelievers."** (Surah Sad: 27) it is wrong to suppose that the universe was created in vain. A philosophy so utterly flawed as materialism and systems based on it were doomed to failure from the very beginning.

Creation is a fact. In this book we will be examining the evidence for this fact. We will see how materialism has collapsed in the face of modern science and also witness how wonderfully and perfectly the universe has been designed and created by Allah.

CHAPTER I

THE CREATION OF THE UNIVERSE FROM NOTHINGNESS

In its standard form, the big bang theory assumes that all parts of the universe began expanding simultaneously. But how could all the different parts of the universe synchronize the beginning of their expansion? Who gave the command?

Andre Linde, Professor of Cosmology[2]

A century ago, the creation of the universe was a concept that astronomers as a rule ignored. The reason was the general acceptance of the idea that the universe existed in infinite time. Examining the universe, scientists supposed that it was just a conglomeration of matter and imagined that it had no beginning. There was no moment of "creation"–a moment when the universe and everything in it came into being.

This idea of "eternal existence" fit in well with European notions stemming from the philosophy of materialism. This philosophy, originally advanced in the world of the ancient Greeks, held that matter was the only thing that existed in the universe and the universe existed in infinite time and will exist endlessly. This philosophy survived in different forms during Roman times but in the Late Roman Empire and Middle Ages, materialism went into decline as a result of the influence of the Catholic church and Christian philosophy. It was after Renaissance that materialism began to gain broad acceptance among European scholars and scientists, largely because of their devotion to ancient Greek philosophy.

The German philosopher Immanuel Kant was the first person to advance the assertion of "the infinite universe" in the New Age. Scientific discoveries, however, invalidated Kant's assertion.

It was Immanuel Kant who, during the European Enlightenment, reasserted and defended materialism. Kant declared that the universe exists for all time and that every probability, however unlikely, should be regarded as possible. Kant's followers continued to defend his idea of an infinite universe along with materialism. By the beginning of 19th century, the idea that the universe had no beginning–that there was never any moment at which it was created–became widely accepted. It was carried into the 20th century through the works of dialectical materialists such as

Karl Marx and Friedrich Engels.

This notion of an infinite universe fit in very well with atheism. It is not hard to see why. To hold that the universe had a beginning could imply that it was created and that, of course requires a creator–that is, Allah. It was much more convenient and safer to circumvent the issue by putting forward the idea that "the universe exists for eternity", even though there was not the slightest scientific basis for making such a claim. Georges Politzer, who espoused and defended this idea in his books published in the early 20th century, was an ardent champion of both Marxism and materialism.

Putting his trust in the validity of the "infinite universe" model, Politzer opposed the idea of creation in his book *Principes Fondamentaux de Philosophie* when he wrote:

> The universe was not a created object, if it were, then it would have to be created instantaneously by God and brought into existence from nothing. To admit creation, one has to admit, in the first place, the existence of a moment when the universe did not exist, and that something came out of nothingness. This is something to which science can not accede. [3]

Politzer supposed that science was on his side in his defense of the idea of an infinite universe. In fact, science was to prove that the universe indeed had a beginning. And just as Politzer himself declared, if there is creation then there must also be a creator.

The Expansion of Universe and the Discovery of the Big Bang

The 1920s were important years in the development of modern astronomy. In 1922, the Russian physicist Alexandra Friedman produced computations showing that the structure of the universe was not static and that even a tiny impulse might be sufficient to cause the whole structure to expand or contract according to Einstein's Theory of Relativity. George Lemaitre was the first to recognize what Friedman's work meant. Based on these computations, the Belgian astronomer Lemaitre declared

Edwin Hubble discovered that the universe was expanding. Eventually he found evidence of the "the Big Bang", a cataclysmic event whose discovery forced scientists to abandon the notion of an infinite and eternal universe.

that the universe had a beginning and that it was expanding as a result of something that had triggered it. He also stated that the rate of radiation could be used as a measure of the aftermath of that "something".

The theoretical musings of these two scientists did not attract much attention and probably would have been ignored except for new observational evidence that rocked the scientific world in 1929. That year the American astronomer Edwin Hubble, working at the California Mount Wilson observatory, made one of the most important discoveries in the history of astronomy. Observing a number of stars through his huge telescope, he discovered that their light was shifted towards the red end of the spectrum and, crucially, that this shift was directly related to the distance of the stars from Earth. This discovery shook the very basis of the universe model held until then.

According to the recognized rules of physics, the spectra of light beams travelling towards the point of observation tend towards violet while the spectra of light beams moving away from the point of observation tend towards red. (Just like the fading of a train's whistle as it moves away from the observer) Hubble's observation showed that according to this law, the heavenly bodies were moving away from us. Before long, Hubble made another important discovery; The stars weren't just racing away from Earth; they were racing away from each other as well. The only conclusion that could be derived from a universe where everything moves away from everything else is that the universe constantly "expands".

Hubble had found observational evidence for something that George Lemaitre had "prophesized" a short while ago and one of the greatest minds of our age had recognized almost fifteen years earlier. In 1915, **Albert Einstein** had concluded that the universe could not be static because of calculations based on his recently-discovered theory of relativity (thus anticipating the conclusions of Friedman and Lemaitre). Shocked by his findings, Einstein added a "cosmological constant" to his equations in order to "make the answer come out right" because astronomers assured him that the universe was static and there was no other way to make his equations match such a model. Years later, Einstein was to admit that his cosmological constant was the biggest mistake of his career.

Hubble's discovery that the universe was expanding led to the emergence of another model that needed no fiddling around with to make the equations work right. If the universe was getting bigger as time advanced, going back in time meant that it was getting smaller; and if one went back far enough, everything would shrink and converge at a single point. The conclusion to be derived from this model was that at some time, all the matter in the universe was compacted in a single point-mass that had **"zero volume"** because of its immense gravitational force. Our universe came into being as the result of the explosion of this point-mass that had zero volume. This explosion has come to be called the **"the Big Bang"** and its existence has repeatedly been confirmed by observational evidence.

There was another truth that the Big Bang pointed to. To say that something has zero volume is tantamount to saying that it is "nothing". **The whole universe was created from this "nothing"**. And furthermore this universe had a beginning, contrary to the view of materialism, which holds that "the universe has existed for eternity".

The "Steady-state" Hypothesis

The Big Bang theory quickly gained wide acceptance in the scientific world due to the clear-cut evidence for it. Nevertheless astronomers who favored materialism and adhered to the idea of an infinite universe that materialism seemingly demanded held out against the Big Bang in their struggle to uphold a fundamental tenet of their ideology. The reason was made clear by the English astronomer Arthur Eddington, who said "Philosophically, the notion of an abrupt beginning to the present order of Nature is repugnant to me".[4]

Another astronomer who opposed the Big Bang theory was Fred Hoyle. Around the middle of the 20th century he came up with a new model, which he called "steady-state", that was an extension of the 19th century's idea of an infinite universe. Accepting the incontrovertible evidence that the universe was expanding, he proposed that the universe was infinite in both dimension and time. According to this model, as the universe expanded new matter was continuously coming into existence by itself in just the right amount to keep the universe in a "steady state". With the sole visible aim of supporting the dogma of "matter existed in infinite time", which is the basis of the materialist philosophy, this theory was totally at variance with the "Big Bang theory", which defends that the universe had a beginning. Supporters of Hoyle's steady state theory remained adamantly opposed to the Big Bang for years. Science, however, was working against them.

The Triumph of the Big Bang

In 1948, George Gamov carried George Lemaitre's calculations several steps fur-

Sir Arthur Eddington's statement that "the notion of an abrupt beginning to the present order of nature was repugnant to him" was an admission of the discomfort that the Big Bang caused for materialists.

ther and came up with a new idea concerning the Big Bang. If the universe was formed in a sudden, cataclysmic explosion, there ought to be a definite amount of radiation left over from that explosion. This radiation should be detectable and, furthermore, it should be uniform throughout the universe.

Within two decades, observational proof of Gamov's conjecture was forthcoming. In 1965, two researchers by the name of Arno Penzias and Robert Wilson chanced upon a form of radiation hitherto unnoticed. Called **"cosmic background radiation"**, it was unlike anything coming from anywhere else in the universe for it was extraordinarily uniform. It was neither localized nor did it have a definite source; instead, it was distributed equally everywhere. It was soon realized that this radiation was the echo of the Big Bang, still reverberating since the first moments of that great explosion. Gamov had been spot-on for the frequency of the radiation was nearly the same value that scientists had predicted it would be. Penzias and Wilson were awarded a Nobel prize for their discovery.

The cosmic background radiation discovered by Penzias and Wilson is regarded as incontrovertible evidence of the Big Bang by the scientific world.

In 1989, George Smoot and his NASA team sent a satellite into space. Called the "Cosmic Background Emission Explorer" (COBE), it took only eight minutes for the sensitive instruments on board the satellite to detect and confirm the levels of radiation reported by Penzias and Wilson. These results conclusively demonstrated the existence of the hot, dense form remaining from the explosion out of which the universe came into being. Most scientists acknowledged that COBE had successfully captured the remnants of the Big Bang.

More evidence for the Big Bang was forthcoming. One piece had to do with the relative amounts of hydrogen and helium in the universe.

Observations indicated that the mix of these two elements in the universe was in accord with theoretical calculations of what should have been remained after the Big Bang. That drove another stake into the heart of the steady state theory because if the universe had existed for eternity and never had a beginning, all of its hydrogen should have been burned into helium.

Confronted by such evidence, the Big Bang gained the near-complete approval of the scientific world. In an article in its October 1994 issue, *Scientific American* noted that the Big Bang model was the only one that could account for the constant expansion of the universe and for other observational results.

Defending the Steady-State theory along with Fred Hoyle, Dennis Sciama describes their predicament in the face of the evidence for the Big Bang saying that he had first taken a stand along with Hoyle but, as evidence began to pile up, he had to admit that the game was over and that the steady-state theory had to be dismissed.[5]

Who Created the Universe From Nothing?

With this triumph of the Big Bang, the thesis of an "infinite universe", which forms the basis of materialist dogma, was tossed onto the scrapheap of history. But for materialists it also raised a couple of inconvenient questions: *What existed before the Big Bang? And what force could have caused the great explosion that resulted in a universe that did not exist before?*

Materialists like Arthur Eddington recognized that the answers to these questions could point to the existence of a supreme creator and that they did not like. The atheist philosopher Anthony Flew commented on this point:

> Notoriously, confession is good for the soul. I will therefore begin by confessing that **the Stratonician atheist has to be embarrassed by the contemporary cosmological consensus**. For it seems that the cosmologists are providing a scientific proof of what St. Thomas contended could not be proved philosophically; namely, that the universe had a beginning. So long as the universe can be comfortably thought of as being not only without end but

also beginning, it remains easy to urge that its brute existence, and whatever are found to be its most fundamental features, should be accepted as the explanatory ultimates. Although I believe that it remains still correct, it certainly is neither easy nor comfortable to maintain this position in the face of the Big Bang story. [6]

Many scientists who do not force themselves to be atheists accept and favor the existence of a creator having an infinite power. For instance, the American astrophysicist Hugh Ross proposes a Creator of universe, Who is above all physical dimensions as:

> By definition, time is that dimension in which cause-and-effect phenomena take place. No time, no cause and effect. If time's beginning is concurrent with the beginning of the universe, as the space-time theorem says, then the cause of the universe must be some entity operating in a time dimension completely independent of and pre-existent to the time dimension of the cosmos. ...It tells us that the Creator is transcendent, operating beyond the dimensional limits of the universe. It tells us that God is not the universe itself, nor is God contained within the universe.[7]

Objections to Creation and Why They are Flawed

It is patently obvious that the Big Bang means the creation of the universe out of nothing and this is surely evidence of willful creation. Regarding this fact, some materialist astronomers and physicists have tried to advance alternative explanations to oppose this reality. Mention has already been made of the steady state theory and it was pointed out it was clung to, by those who were uncomfortable with the notion of "creation from nothingness", despite all the evidence to the contrary in an attempt to shore up their philosophy.

There are also a number of models that have been advanced by materialists who accept the Big Bang theory but try to exorcise it of the notion of creation. One of these is the "oscillating" universe model; another is the "quantum model of universe". Let us examine these theories and see why they are invalid.

The oscillating universe model was advanced by the astronomers who disliked the idea the Big Bang was the beginning of the universe. In this model, it is claimed that the present expansion of the universe will eventually be reversed at some point and begin to contract. This contraction will cause everything to collapse into a single point that will then explode again, initiating a new round of expansion. This process, they say, is repeated infinitely in time. This model also holds that the universe has experienced this transformation an infinite number of times already and that it will continue to do so forever. In other words, the universe exists for eternity but it expands and collapses at different intervals with a huge explosion punctuating each cycle. The universe we live in is just one of those infinite universes going through the same cycle.

This is nothing but a feeble attempt to accommodate the fact of the Big Bang to notions about an infinite universe. The proposed scenario is unsupported by the results of scientific research over the last 15-20 years, which show that it is impossible for such an "oscillating" universe idea to come into being. Furthermore the laws of physics offer no reason why a contracting universe should explode again after collapsing into a single point: it ought to stay just as it is. Nor do they offer a reason why an expanding universe should ever begin to contract in the first place.[8]

Even if we allow that there is some mechanism by which this cycle of contraction-explosion-expansion does take place, the crucial point is that this cycle cannot go on for ever, as is claimed. Calculations for this model show that each universe will transfer an amount of entropy to its successor. In other words, the amount of useful energy available becomes less each time and every "opening" universe will open more slowly and have a larger diameter. This will cause a much smaller universe to form the next time around and so on, eventually petering out into nothing. Even if "open and close" universes can exist, they cannot endure for eternity. At some point it becomes necessary for "something" to be created from "nothing".[9]

Put briefly, the "oscillating" universe model is a hopeless fantasy whose physical reality is impossible.

The **"quantum model of universe"** is another attempt to purge the

Big Bang of its creationist implications. Supporters of this model base it on the observations of quantum (subatomic) physics. In quantum physics, it is to be observed that subatomic particles appear and disappear spontaneously in a vacuum. Interpreting this observation as "matter can originate at quantum level, this is a property pertaining to matter", some physicists try to explain the origination of matter from non-existence during the creation of the universe as a "property pertaining to matter" and present it as a part of laws of nature. In this model, our universe is interpreted as a subatomic particle in a bigger one.

However this syllogism is definitely out of question and in any case cannot explain how the universe came into being. William Lane Craig, the author of *The Big Bang: Theism and Atheism* explains why:

> A quantum mechanical vacuum spawning material particles is far from the ordinary idea of a "vacuum" (meaning nothing). Rather, a quantum vacuum is a sea of continually forming and dissolving particles, which borrow energy from the vacuum for their brief existence. This is not "nothing," and hence, material particles do not come into being out of nothing.[10]

So in quantum physics, matter "does not exist when it was not before". What happens is that ambient energy suddenly becomes matter and just as suddenly disappears becoming energy again. In short, there is no condition of "existence from nothingness" as is claimed.

In physics, no less than in other branches of the sciences, there are atheist scientists who do not hesitate to disguise the truth by overlooking critical points and details in their attempt to support the materialist view and achieve their ends. For them, it is much more important to defend materialism and atheism than to reveal scientific facts and realities.

In the face of the reality mentioned above, most scientists dismiss the quantum universe model. C. J. Isham explains that "this model is not accepted widely because of the inherent difficulties that it poses."[11] Even some of the originators of this idea, such as Brout and Spindel, have abandoned it.[12]

A recent and much-publicized version of the quantum universe model was advanced by the physicist **Stephen Hawking**. In his book *A*

Brief History of Time, Hawking states that the Big Bang doesn't necessarily mean existence from nothingness. Instead of "no time" before the Big Bang, Hawking proposed the concept of **"imaginary time"**. According to Hawking, there was only a 10^{-43} second "imaginary" time interval before the Big Bang took place and "real" time was formed after that. Hawking's hope was just to ignore the reality of "timelessness" before the Big Bang by means of this "imaginary" time.

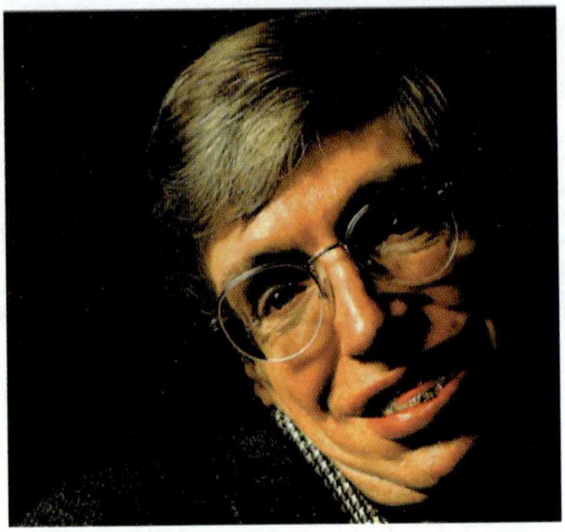

Stephen Hawking also tries to advance different explanations for the Big Bang other than Creation just as other Materialist scientists do by relying upon contradictions and false concepts.

As a concept, "imaginary time" is tantamount to zero or non-existence–like the imaginary number of people in a room or the imaginary number of cars on a road. Here Hawking is just playing with words. He claims that equations are right when they are related to an imaginary time but in fact this has no meaning. The mathematician Sir Herbert Dingle refers to the possibility of faking imaginary things as real in math as:

> In the language of mathematics we can tell lies as well as truths, and within the scope of mathematics itself there is no possible way of telling one from the other. We can distinguish them only by experience or by reasoning outside the mathematics, applied to the possible relation between the mathematical solution and its physical correlate.[13]

To put it briefly, a mathematically imaginary or theoretical solution need not have a true or a real consequence. Using a property exclusive to mathematics, Hawking produces hypotheses that are unrelated to reality. But what reason could he have for doing this? It's easy to find the answer to that question in his own words. Hawking admits that he prefers alternative universe models to the Big Bang because the latter "hints at divine creation", which such models are designed to oppose.[14]

What all this shows is that alternative models to the Big Bang such as steady-state, the open and close universe model, and quantum universe models in fact spring from the philosophical prejudices of materialists. Scientific discoveries have demonstrated the reality of the Big Bang and can even explain "existence from nothingness". And this is very strong evidence that the universe is created by Allah, a point that materialists utterly reject.

An example of this opposition to the Big Bang is to be found in an essay by John Maddox, the editor of *Nature* (a materialist magazine), that appeared in 1989. In ***"Down with the Big Bang"***, Maddox declares the Big Bang to be philosophically unacceptable because **it helps theologists by providing them with strong support for their ideas.** The author also predicted that the Big Bang would be disproved and that support for it would disappear within a decade.[15] Maddox can only have been even more discomforted by the subsequent discoveries during the next ten years that have provided further evidence of the existence of the Big Bang.

Some materialists do act with more common sense on this subject. The British Materialist H. P. Lipson accepts the truth of creation, albeit "unpleasantly", when he says:

> If living matter is not, then caused by the interplay of atoms, natural forces, and radiation, how has it come into being?...I think, however, that **we must...admit that the only acceptable explanation is creation**. I know that this is anathema to physicists, as indeed it is to me, but we must not reject that we do not like if the experimental evidence supports it.[16]

In conclusion, the truth disclosed by science is this: Matter and time have been brought into being by an independent possessor of immense power, by a Creator. Allah, the Possessor of almighty power, knowledge and intelligence, has created the universe we live in.

The Signs of the Qur'an

In addition to explaining the universe, the Big Bang model has another important implication. As the quotation from Anthony Flew cited above points out, science has proven an assertion hitherto supported only by religious sources.

The truth that is defended by religious sources is the reality of creation from nothingness. This has been declared in the holy books that have served as guides for mankind for thousands of years. In all holy books such as the Old Testament, New Testament, and the Qur'an, it is declared that the universe and everything in it were created from nothingness by Allah.

In the only book revealed by Allah that has survived completely intact, the Qur'an, there are statements about the creation of the universe from nothing as well as how this came about that are parallel to 20th-century knowledge and yet were revealed fourteen centuries ago.

First of all, the creation of this universe from nothingness is revealed in the Qur'an as follows:

He(Allah) is the Originator of the heavens and the earth…(Surat al-Anam: 101)

Another important aspect revealed in the Qur'an fourteen centuries before the modern discovery of the Big Bang and findings related to it is that when it was created, the universe occupied a very tiny volume:

Do those who are disbelievers not see that the heavens and the earth were sewn together and then We unstitched them and that We made from water every living thing? So will they not have faith? (Surat al-Anbiya': 30)

There is a very important choice of words in the original Arabic whose translation is given above. The word *ratk* translated as **"sewn to"** means "mixed in each, blended" in Arabic dictionaries. It is used to refer to two different substances that make up a whole. The phrase **"we unstitched"** is the verb *fatk* in Arabic and implies that something comes into being by tearing apart or destroying the structure of *ratk*. The sprouting of a seed from the soil is one of the actions to which this verb is applied.

Let us take a look at the verse again with this knowledge in mind. In the verse, sky and earth are at first subject to the status of *ratk*. They are separated (*fatk*) with one coming out of the other. Intriguingly, cosmologists speak of a "cosmic egg" that consisted of all the matter in the universe prior to the Big Bang. In other words, all the heavens and earth were included in this egg in a condition of *ratk*. This cosmic egg exploded violently causing its matter to *fatk* and in the process created the structure of the whole universe.

Another truth revealed in the Qur'an is the expansion of the universe that was discovered in the late 1920s. Hubble's discovery of the red shift in the spectrum of starlight is revealed in the Qur'an as :

It is We Who have built the universe with (Our creative) power, and, verily, it is We Who are steadily expanding it. (Surat adh-Dhariyat: 47)

In short, the findings of modern science support the truth that is revealed in the Qur'an and not materialist dogma. Materialists may claim this all to be "coincidence" but the plain fact is that the universe came into being as a result of an act of creation on the part of Allah and the only true knowledge about the origin of universe is to be found in the word of Allah as revealed to us.

CHAPTER II

THE EQUILIBRIUM IN THE EXPLOSION

The explosive vigour of the universe is thus matched with almost unbelievable accuracy to its gravitating power. The big bang was not evidently, any old bang, but an explosion of exquisitely arranged magnitude.

Paul Davies, Professor of Theoretical Physics[17]

In the first chapter we examined the universe's creation from nothingness as a result of a great explosion. Let us now consider some of the implications of this.

Scientists estimate that there are over 300 billion galaxies in the whole universe. These galaxies have a number of different forms (spiral, elliptical, etc) and each contains about as many stars as the universe contains galaxies. One of these stars, the Sun, has nine major planets rotating around in it in great harmony. All of us live on the third of those planets counting from the sun.

Look about you: Does what you see appear to be a disordered jumble of matter haphazardly scattered this way and that? Of course not. But how could matter have formed organized galaxies if it had been dispersed randomly? Why has matter accumulated at certain points and formed stars? How could the delicate balance of our solar system have emerged from a violent explosion? These are very important questions and they lead us to the real question of how the universe was structured after the Big Bang.

If the Big Bang was indeed a such cataclysmic explosion then it is reasonable to expect that matter should have been scattered everywhere at random. And yet it is not. Instead it is organized into planets, and stars, and galaxies, and clusters of galaxies, and superclusters of galaxies. It is as if a bomb that exploded in a granary caused all the wheat to fall into neat sacks and bales on the backs of trucks ready to be delivered instead of showering the grains every which way. Fred Hoyle, a staunch opponent of the Big Bang theory for years, expressed his own surprise at this structure:

> The big bang theory holds that the universe began with a single explosion. Yet as can be seen below, an explosion merely throws matter apart, while **the big bang has mysteriously** produced the opposite effect- **with matter clumping together in the form of galaxies.**[18]

That the matter produced by the Big Bang should have formed such tidy and organized shapes is indeed an extraordinary thing. The occurrence of such a harmony leads us to the realization that the universe was the result of its perfect creation by Allah.

In this chapter we will examine and consider this extraordinary perfection and excellence.

The Speed of the Explosion

People hearing of the Big Bang but not considering the subject at length do not think about what an extraordinary plan must lie behind this explosion. That's because the notion of an explosion doesn't suggest harmony, plan, or organization to most people. In fact there are a number of very puzzling aspects to the intricate order in the Big Bang.

One of these puzzles has to do with the acceleration caused by the explosion. When the explosion took place, matter certainly must have begun moving at an enormous speed in every direction. But there is another point that we need to pay attention to here. There must also have been a very big attractive force at the first moment of the explosion: an attractive force that was strong enough to gather the whole universe into one point.

Two different and opposing forces are at work here. The force of the explosion, driving matter outward and away, and the force of attraction, trying to resist the first and pull everything back together. The universe came into being because these two forces were in equilibrium. If the attractive force had been greater than the explosive, the universe would have collapsed. If the opposite had been true, matter would have been splattered in every direction in a way never to unite again.

Then how sensitive was this equilibrium? How much "slack" could there have been between the two forces?

The mathematical physicist Paul Davies, a professor at the University of Adelaide in Australia, performed lengthy calculations of the conditions that must have existed at the moment of the Big Bang and came up with a result that can only be described as astonishing. According to Davies, **if the rate of expansion had differed by more than 10^{-18} seconds (one quintillionth of a second)**, there would have been no universe. Davies describes his conclusion:

Paul Davies: "The evidence is strong enough to acknowledge the existence of a conscious cosmic design."

> Careful measurements puts the rate of expansion very close to a critical value at which the universe will just escape its own gravity and expand forever. A little slower and the cosmos would collapse, a little faster and the cosmic material would have long ago completely dispersed. It is interesting to ask precisely how delicately the rate of expansion has been "fine tuned" to fall on this narrow dividing line between two catastrophes. If at time I S (by which the time pattern of expansion was already firmly established) the expansion rate had differed from its actual value by more than 10^{-18}, it would have been sufficient to throw the delicate balance out. The explosive vigour of the universe is thus matched with almost unbelievable accuracy to its gravitating power. **The big bang was not evidently, any old bang, but an explosion of exquisitely arranged magnitude.**[19]

Bilim Teknik (Science Technique, a Turkish scientific periodical) quotes an article that appeared in *Science* in which the phenomenal equilibrium that obtained in the initial phase of universe is stated:

> If the density of the universe was a little bit more, in that case, according to Einstein's relativity theory, the universe would not be expanding due to the attraction forces of atomic particles but contracting, ultimately diminishing to a spot. If the initial density had been a little bit less, then the universe would rapidly be expanding, but in this case, atomic particles would not be attracting each other and no stars and no galaxies would ever have formed. Consequently, man would never come into existence! According to the calculations, the difference between the initial real density of the universe and its critical density, which is unlikely to occur, is less than one percent's one quadrillion. **This is similar to place a pencil in a position so that it can stand on its sharp end even after one billion years...** Furthermore, as the universe expands, this equilibrium becomes more delicate.[20]

It is We Who have built the universe with (Our creative) power, and, verily, it is We Who are steadily expanding it. (Surat adh-Dhariyat: 47)

Even Stephen Hawking, who tries hard to explain away the creation of the universe as a series coincidences in *A Brief History of Time*, acknowledges the extraordinary equilibrium in the rate of expansion:

> If the rate of expansion one second after the big bang had been smaller by even one part in a hundred thousand million million, the universe would have recollapsed before it ever reached its present size. [21]

What then does such a remarkable equilibrium as this indicate? The only rational answer to that question is that it is proof of conscious design and cannot possibly be accidental. Despite his own materialist bent, Dr Davies admits this himself:

> It is hard to resist that the present structure of the universe, apparently so sensitive to minor alterations in the numbers, has been rather carefully thought out... The seemingly miraculous concurrence of numerical values that nature has assigned to her fundamental constants must remain the most compelling evidence for an element of **cosmic design**.[22]

The Four Forces

The speed of the Big Bang's explosion is only one of the remarkable states of equilibrium at the initial moment of creation. Immediately after the Big Bang, forces that underpin and organize the universe we live in had to be numerically "just right" otherwise there would have been no universe.

These are the "four fundamental forces" that are recognized by modern physics. All structure and motion in the universe is governed by these four forces, known as the gravitational force, the electromagnetic force, the strong nuclear force, and the weak nuclear force. The strong and weak nuclear forces operate only at the atomic scale. The remaining two–the gravitational force and the electromagnetic force–govern assemblages of atoms, in other words "matter". These four fundamental forces were at work in the immediate aftermath of the Big Bang and resulted in the creation of atoms and matter.

A comparison of those forces is enlightening for their values are stunningly different from one another. Below they are given in international standard units:

Strong nuclear force:	15
Weak nuclear force:	7.03×10^{-3}
Electromagnetic force:	3.05×10^{-12}
Gravitational force:	5.90×10^{-39}

Notice how great are the differences in the strengths of these four fundamental forces. The difference between the strongest (strong nuclear force) and the weakest (gravitational force) is about 25 followed by 38 zeros! Why should this be so?

The molecular biologist Michael Denton addresses this question in his book, *Nature's Destiny*:

> If, for example, the gravitational force was a trillion times stronger, then the universe would be far smaller and its life history far shorter. An average star would have a mass a trillion times less than the sun and a life span of about one year. On the other hand, if gravity had been less powerful, no stars or galaxies would have ever formed. The other relationships and values are no less critical. If the strong force had been just slightly weaker, the only element that would be stable would be hydrogen. No other atoms could exist. If it had been slightly stronger in relation to electromagnetism, then an atomic nucleus consisting of only two protons would be a stable feature of the universe-which would mean there would be no hydrogen, and if any stars or galaxies evolved, they would be very different from the way they are. Clearly, **if these various forces and constants did not have precisely the values they do, there would be no stars, no supernovae, no planets, no atoms, no life.** [23]

Paul Davies comments on how the laws of physics provide for conditions ideal

The molecular biologist Michael Denton addresses an important point in his book, Nature's Destiny: How the Laws of Biology Reveal Purpose in the Universe. According to Denton, the universe was created and specially designed to make human life possible.

eople to live:

Had nature opted for a slightly different set of numbers, the world would be a very different place. Probably we would not be here to see it...Recent discoveries about the primeval cosmos oblige us to accept that **the expanding universe has been set up in its motion with a cooperation of astonishing precision.**[24]

Arno Penzias, who was the first, along with Robert Wilson to detect the cosmic background radiation (for which discovery the pair received a Nobel prize in 1965), comments on the beautiful design in the universe:

Astronomy leads us to a unique event, a universe which was created out of nothing, one with the very delicate balance needed to provide exactly the conditions required to permit life, and one which has underlying (one might say "supernational") plan.[25]

The scientists we have just quoted have all drawn an important conclusion from their observations. Examining and thinking about the incredible balances and their beautiful order in the design of universe inevitably leads one to a truth: There exists in this universe a superior design and a perfect harmony. Unquestionably the Author of this design and harmony is Allah, Who has created everything flawlessly. Allah draws our attention in one of His verses to the order in the creation of the universe, planned, and computed in every detail:

He to whom the kingdom of the heavens and the earth belongs. He does not have a son and He has no partner in the Kingdom. He created everything and determined it most exactly. (Surat al-Furqan: 2)

The Mathematics of Probability Refutes "Coincidence"

What has been said so far shows the extraordinary balances among the forces that make human life possible in this universe. The speed of the Big Bang's explosion, the values of the four fundamental forces, and all the other variables that we will be examining in the chapters ahead and which are vital for existence have been arranged according to an extraordinary precision.

Let us now make a brief digression and consider the coincidence the-

THE PROBABILITY OF THE OCCURRENCE OF A UNIVERSE IN WHICH LIFE CAN FORM

The calculations of British mathematician Roger Penrose show that the probability of universe conducive to life occurring by chance is 1 in $10^{10^{123}}$. The phrase "extremely unlikely" is inadequate to describe this possibility.

10

1000000000000000000000000000000
0000000000000000000000000000000
0000000000000000000000000000000
0000000000000000000000000000000

ory of materialism. Coincidence is a mathematical term and the possibility of an event's occurrence can be calculated using the mathematics of probability. Let's do so.

Taking the physical variables into account, what is the likelihood of a universe giving us life coming into existence by coincidence? One in billions of billions? Or trillions of trillions of trillions? Or more?

Roger Penrose, a famous British mathematician and a close friend of Stephen Hawking, wondered about this question and tried to calculate the probability. Including what he considered to be all variables required for human beings to exist and live on a planet such as ours, he computed the probability of this environment occurring among all the possible results of the Big Bang.

According to Penrose, the odds against such an occurrence were on the order of $10^{10^{123}}$ to 1.

It is hard even to imagine what this number means. In math, the value 10^{123} means 1 followed by 123 zeros. (This is, by the way, more than the total number of atoms 10^{78} believed to exist in the whole universe.) But Penrose's answer is vastly more than this: It requires 1 followed by 10^{123} zeros.

Roger Penrose: "This number tells us how precise the Creator's aim must have been."

Or consider: 10^3 means 1,000, a thousand. 10^{10^3} is a number that that has 1 followed by 1000 zeros. If there are six zeros, it's called a million; if nine, a billion; if twelve, a trillion and so on. There is not even a name for a number that has 1 followed by 10^{123} zeros.

In practical terms, in mathematics, a probability of 1 in 10^{50} means "zero probability". Penrose's number is more than trillion trillion trillion times less than that. In short, Penrose's number tells us that the 'accidental" or "coincidental" creation of our universe is an impossibility.

Concerning this mind-boggling number Roger Penrose comments:

> This now tells how precise the Creator's aim must have been, namely to an accuracy of one part in $10^{10^{123}}$. This is an extraordinary figure. One could not possibly even write the number down in full in the ordinary denary notation: it would be 1 followed by 10^{123} successive 0's. Even if we were to write a 0 on each separate proton and on each separate neutron in the entire universe- and we could throw in all the other particles for good measure- we should fall far short of writing down the figure needed. [26]

The numbers defining the design and plan of the universe's equilibrium play a crucial role and exceed comprehension. They prove that the universe is by no means the product of a coincidence, and show us "how precise the Creator's aim must have been" as Penrose stated.

In fact in order to recognize that the universe is not a "product of coincidences" one does not really need any of these calculations at all. Simply by looking around himself, a person can easily perceive the fact of creation in even the tiniest details of what he sees. How could a universe like this, perfect in its systems, the sun, the earth, people, houses, cars, trees, flowers, insects, and all the other things in it ever have come into existence as the result of atoms falling together by chance after an explosion? Every detail we peer at shows the evidence of Allah's existence and supreme power. Only people who reflect can grasp these signs.

> **In the creation of the heavens and earth, and the alternation of the night and day, and the ships which sail the seas to people's benefit, and the water which Allah sends down from the sky- by which He brings the earth to life when it was dead and scatters about in it creatures of every kind - and the varying direction of the winds, and the clouds subservient between heaven and earth, there are Signs for people who use their intellect. (Surat al-Baqara:164)**

Seeing the Plain Truth

20th-century science has come up with categorical evidence that the universe was created by Allah. The anthropic principle that we mentioned before reveals every detail of a universe that has been designed for mankind to live in and in which there is no place for chance.

The interesting part is that the ones who discovered all this and came to the conclusion that the universe couldn't possibly have come into being by accident are the very same people who defend the philosophy of materialism. Scientists such as Paul Davies, Arno Penzias, Fred Hoyle and Roger Penrose are not pious men and they certainly had no intention of proving Allah's existence as they pursued their work. One can imagine that they reached their conclusions about the design of the universe by a superior power most unwillingly.

The American astronomer George Greenstein confesses this in his book *The Symbiotic Universe*:

> How could this possibly have come to pass (that the laws of physics conform themselves to life)?...As we survey all the evidence, the thought insistently arises that some supernatural agency - or, rather Agency- must be involved. Is it possible that suddenly, without intending to, we have stumbled upon scientific proof of the existence of a Supreme Being? Was it God who

stepped in and so providentially crafted the cosmos for our benefit?[27]

An atheist, Greenstein disregards the plain truth; nevertheless he cannot keep from wondering. Other, less prejudiced scientists on the other hand, readily admit that the universe must have been specially designed for mankind to live in. The American astrophysicist Hugh Ross ends his article *"Design and the Anthropic Principle"* with these words:

> An intelligent, transcendent Creator must have brought the universe into existence. An intelligent, transcendent Creator must have designed the universe. An intelligent, transcendent Creator must have designed planet Earth. An intelligent, transcendent Creator must have designed life.[28]

Thus science proves the reality of creation. Certainly there is Allah and He has created everything around us–the seen and the unseen. He is the sole Creator of the extraordinary and outstanding equilibrium and design of the heavens and Earth.

It has come such a pass that today, materialism has the flavor of a superstitious, unscientific system of belief. The American geneticist Robert Griffiths jokingly remarked "If we need an atheist for a debate, I go to the philosophy department. The physics department isn't much use."[29]

To sum up: Every physical law and every physical constant in this universe has been specifically designed to enable human beings to exist and live. In his book *The Cosmic Blueprint,* Davies states this truth in the last paragraph, "The impression of Design is overwhelming."[30]

Doubtlessly, the design of the universe is evidence of Allah's power to establish. The precise balances and all the human beings and other creatures are the evidence of Allah's supreme power and act of creation. This result discovered by modern science is just a reworking of a truth revealed fourteen centuries ago in the Qur'an:

> **Your Lord is Allah, Who created the heavens and the earth in six days and then settled Himself firmly on the Throne. He covers the day with the night and, each pursuing the other urgently; and the sun and moon and stars are subservient to His command. Both creation and command belong to Him. Blessed be Allah, the Lord of all worlds. (Surat al-Araf: 54)**

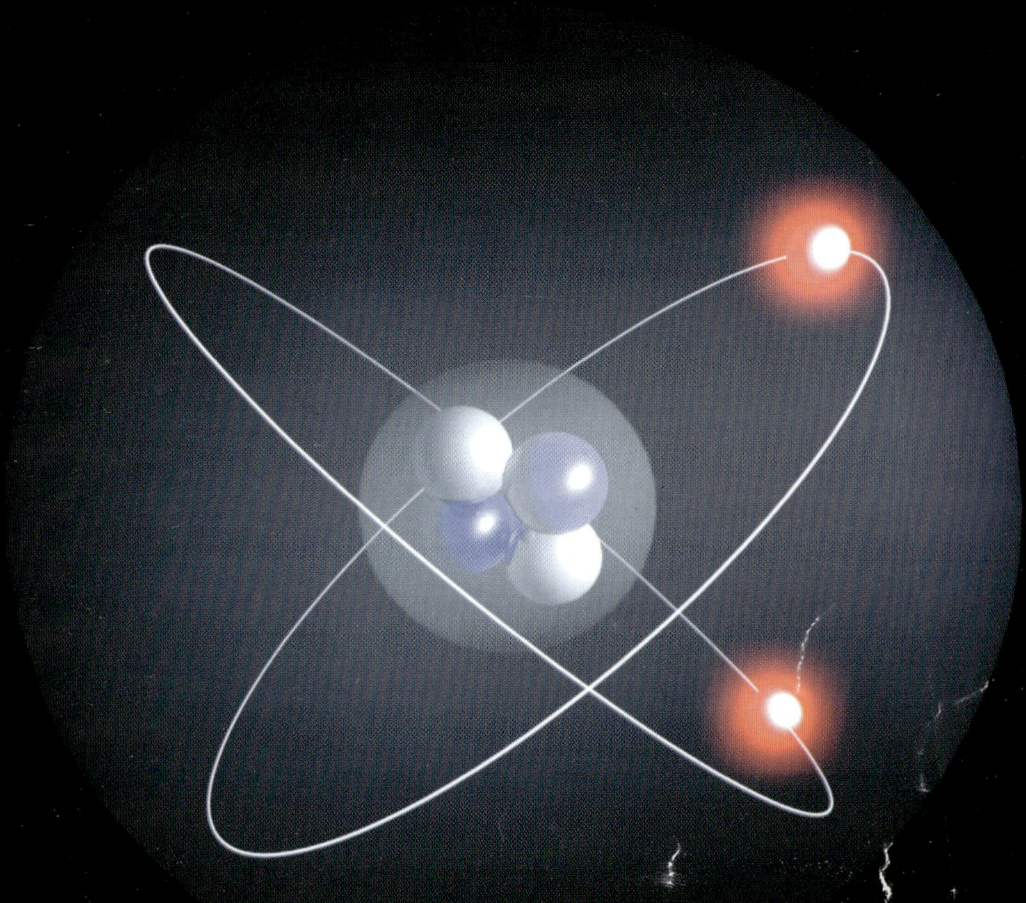

CHAPTER III

THE RHYTHM OF THE ATOMS

If the world's finest minds can unravel only with difficulty the deeper workings of nature, how could it be supposed that those workings are merely a mindless accident, a product of blind chance?

Paul Davies, Professor of Theoretical Physics [31]

Scientists are in general agreement that, on the basis of calculations, the Big Bang took place about 17 billion years ago. All the matter making up the universe was created from nothingness but with the wonderful design that we talked about in the first two chapters. Nevertheless, the universe that emerged from the Big Bang could have been much different from the one that did emerge–ours.

For example, if the values of four fundamental forces were different, the universe would have consisted of only radiation and become a tissue of light with no stars, galaxies, human beings, or anything else. Thanks to the extraordinary perfect balance of those four forces, "atoms"–the building-blocks of that which is called "matter"–came into being.

Scientists are also in general agreement that the first two simplest elements–hydrogen and helium–began to form during the first fourteen seconds after the Big Bang. The elements were formed as a result of a reduction in the universal entropy that was causing matter to scatter everywhere. In other words, at first the universe was just an amassing of hydrogen and helium atoms. If it had remained so, again there could have been no stars, planets, stones, soil, trees, or human beings. It would have been a lifeless universe consisting of only those two elements.

Carbon, the fundamental element of life, is a much heavier element than hydrogen and helium. How did it come into being?

Searching for an answer to this question, scientists stumbled upon one of the most surprising discoveries of this century.

The Structure of the Elements

Chemistry is a science that deals with the composition, structure, and properties of substances and with the transformations that they undergo. The bedrock of modern chemistry is the periodic table of elements. First laid out by Russian chemist Dmitry Ivanovich Mendeleyev, the elements in the periodic table are arranged according to their atomic structures. Hydrogen occupies the first place in the table because it is the simplest of all the elements, consisting of only one proton in its nucleus and one electron revolving around it.

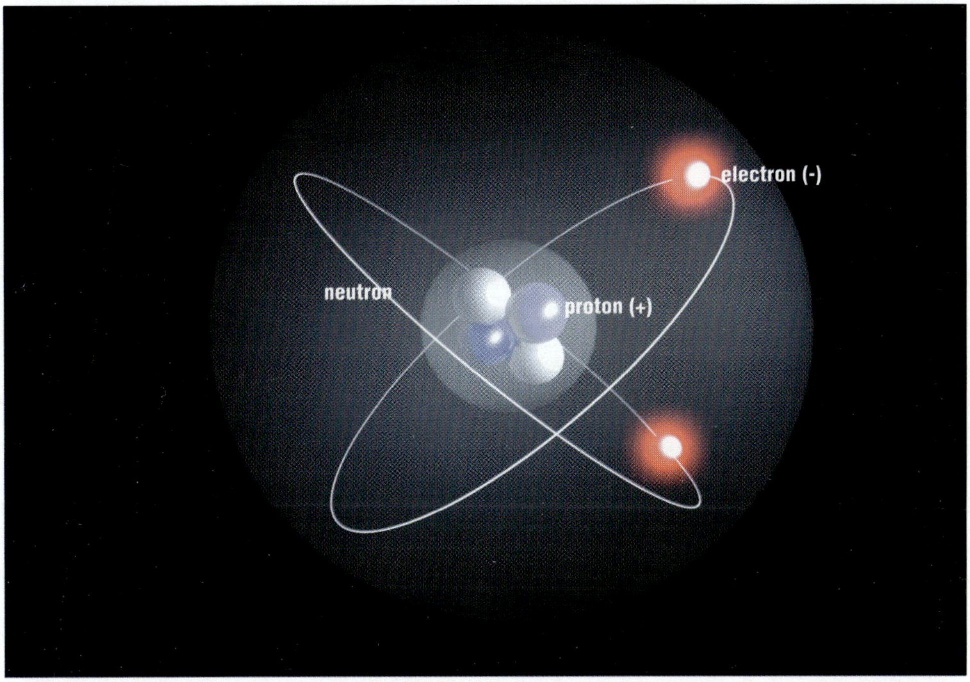

Protons are subatomic particles that carry a positive electrical charge in the nucleus of an atom. Helium, with two protons, occupies the second place in the periodic table. Carbon has six protons and oxygen has eight. All the elements differ in the number of protons that they contain.

Another particle present in the nucleus of an atom is the neutron. Unlike protons, neutrons do not carry an electrical charge: they are neutral in other words, hence their name.

The third basic particle of which atoms are composed is the electron, which has a negative electrical charge. In every atom, the number of protons and electrons is the same. Unlike protons and neutrons however, electrons are not located in the nucleus. Instead, they move around the nucleus at a very high speed that keeps the positive and negative charges of the atom apart.

The differences in atomic structure (the numbers of protons/electrons) are what make the elements different from one another.

A crucial rule of (classical) chemistry is that elements cannot be transformed into one another. Changing iron (with twenty-six protons) into silver (with eighteen) would require removing eight protons from the nucleus. But protons are bound together by the strong nuclear force and the number of protons in a nucleus can be changed only in nuclear reactions. Yet all the reactions that take place under terrestrial conditions are chemical reactions that depend on electron exchange and that do not effect the nucleus.

In the Middle Ages there was a "science" called alchemy–the forerunner of modern chemistry. Alchemists, unaware of the periodic table or the atomic structures of the elements, thought it was possible to transform one element into another. (A favorite object of pursuit, for reasons that should be apparent, was trying to turn iron into gold.) We now know that what the alchemists were trying to do is impossible under normal conditions such as exist on Earth: The temperatures and pressures required for such a transformation to take place are too enormous to achieve in any terrestrial laboratory. But it is possible if you have the right place to do it in.

And the right place, it turns out, is in the hearts of stars.

The Universe's Alchemy Labs: Red Giants

The temperature required to overcome the reluctance of nuclei to change is nearly 10 million degrees Celsius. This is why "alchemy" in the real sense takes place only in stars. In medium-sized stars like the Sun, the enormous energy being radiated is the result of hydrogen being fused into helium.

Keeping this brief review of the chemistry of elements in mind, let us return to the immediate aftermath of the Big Bang. We mentioned that only helium and hydrogen atoms existed in the universe after the Big Bang. Astronomers believe that solar-type stars (of which the Sun is one) are formed as a result of nebulae (clouds) of hydrogen and helium gas being compressed until the hydrogen-to-helium thermonuclear reaction gets started. So now we have stars. But our universe is still lifeless. For life, heavier elements–oxygen and carbon specifically–are required. There needs to be another process whereby hydrogen and helium can be con-

Red giants are huge stars about fifty times bigger than our sun. Deep within these giants, an extraordinary process takes place.

Helium nucleus

Carbon nucleus

verted into still other elements.

The "manufacturing-plants" of these heavy elements it turns out are the red giants–a class of stars that are fifty times bigger than the Sun.

Red giants are much hotter than solar-type stars and this characteristic enables them to do something other stars cannot: They convert helium into carbon. Nevertheless, even for a red giant this is not easy. As the astronomer Greenstein says: **"Even now, when the answer (as to how they do it) is well in hand, the method they employ seems astonishing."**[32]

Helium's atomic weight is 2: that is, it has two protons in its nucleus. Carbon's atomic weight is 6. In the fantastically high temperatures of red giants, three helium atoms are fused into a carbon atom. This is the "alchemy" that supplied the universe with its heavier elements after the Big Bang.

But as we said: it's not easy. It's nearly impossible to persuade two helium atoms to join together and quite impossible for three. So how do the six protons needed for carbon get together?

It's a two-step process. First, two helium atoms are fused into an intermediary element with four protons and four neutrons. Next, a third helium is added to this intermediary element to make a carbon atom with six protons and six neutrons.

The extraordinarily unstable isotope of beryllium that is formed in red giants.

Normal beryllium as found on Earth.

The intermediary element is **beryllium**. Beryllium occurs naturally on Earth but the beryllium that occurs in red giants is different in a crucially important way: It consists of four protons and four neutrons, whereas terrestrial beryllium has five neutrons. "Red-giant beryllium" is a slightly different version. It's what's called an "isotope" in chemistry.

Now comes the real surprise. The "red-giant" isotope beryllium turns out to be incredibly unstable. Scientists have studied this isotope for years and discovered that once it has formed, **it breaks down again in just 0.000000000000001 second.**

How is this unstable beryllium isotope, which forms and disintegrates in such a short time, able to unite with a helium atom to become a carbon atom? It is like trying to lay a third brick on two other bricks that shoot away from each other in 0.000000000000001 second if they chance to come atop one another, and form a construction in this way. How does this process take place in red giants? Physicists scratched their heads over this puzzle for decades without coming up with an answer. The American astrophysicist Edwin Salpeter finally discovered a clue to the mystery in the concept of "atomic resonance".

Resonance and Double Resonance

Resonance is defined as the harmony of frequencies (vibrations) of two different materials.

A simple example from ordinary experience will give us an idea of what physicists mean by "atomic resonance". Imagine yourself and a child at a playground where there are swings. The child sits on the swing and you give him a push to get him started. To keep the swing moving, you have to keep pushing it from behind. But the timing of these pushes is important. Each time the swing approaches you, you have to apply the force of the push just at the right moment: when the swing is at the highest point of its motion towards you. If you push too soon, the result is a collision that disturbs the rhythmic momentum of the swing; if you push too late, the effort is wasted because the swing is already moving away from you. In other words, the frequency of your pushes must be in harmony with the frequency of the swing's approaches to you.

Physicists refer to such a "harmony of frequencies" as "resonance". The swing has a frequency: for example it reaches you every 1.7 seconds. Using your arms you push it every 1.7 seconds. Of course if you want, you can change the frequency of the swing's motion, but if you do, you have to change the frequency of the pushes as well, otherwise the swing will not swing right.[33]

Just as two or more moving bodies can resonate, resonance can also occur when one moving body causes motion in another. This type of resonance is often seen in musical instruments and is called "acoustic resonance". It can occur, for example, among two finely-tuned violins. If one of these violins is played in the same room as the other, the strings of the second will vibrate and produce a sound even though nobody is touching it. Because both instruments have been precisely tuned to the same frequency, a vibration in one causes a vibration in the other.[34]

The resonances in these two examples are simple ones and are easy to keep the track of. There are other resonances in physics that are not simple at all and in the case of atomic nuclei, the resonances can be quite complex and sensitive.

Every atomic nucleus has a natural energy level that physicists have been able to identify after lengthy study. These energy levels are quite different from one another but a few rare instances of resonance between atomic nuclei have been observed. When such resonance occurs, the motions of the nuclei are in harmony with one another like our examples of the swing and violin. The important point of this is that the resonance expedites nuclear reactions that can affect the nuclei.[35]

Investigating how carbon was made by red giants, Edwin Salpeter suggested that there must be a resonance between helium and beryllium nuclei that facilitated the reaction. This resonance, he said, made it easier for helium atoms to fuse into beryllium and this could account for the reaction in red giants. Subsequent research however failed to support this idea.

Fred Hoyle was the second astronomer to address this question. Hoyle took Salpeter's idea a step further, introducing the idea of "**double resonance**". Hoyle said that there had to be two resonances: one that caused two heliums to fuse into beryllium and one that caused the third helium atom join this unstable formation. Nobody believed Hoyle. The idea of such a precise resonance occurring once was hard enough to accept; that it should occur twice was unthinkable. Hoyle pursued his research for years and in the end he proved that his idea was right: there really was a double resonance taking place in the red giants. At the exact moment two helium atoms resonated in union, a beryllium atom appeared in the 0.000000000000001 second needed to pro-

Fred Hoyle was the first to discover the amazing equilibrium of nuclear reactions taking place in red giants. Although an atheist, Hoyle admitted that this balance could not be explained by chance and that it was a deliberate arrangement.

duce carbon. George Greenstein describes why this double resonance is indeed an extraordinary mechanism:

> There are three quite separate structures in this story-helium, beryllium, and carbon-and two quite separate resonances. It is hard to see why these nuclei should work together so smoothly...Other nuclear reactions do not proceed by such a remarkable chain of lucky breaks...It is like discovering deep and complex resonances between a car, a bicycle, and a truck. **Why should such disparate structures mesh together so perfectly?** Upon this our existence, and that of every life form in the universe, depends.[36]

In the years that followed it was discovered that other elements like oxygen are also formed as a result of such amazing resonances. A zealous materialist, Fred Hoyle's discovery of these "extraordinary transactions" forced him to admit in his book *Galaxies, Nuclei and Quasars*, that such double resonances had to be the result of design and not coincidence. [37] In another article he wrote:

> If you wanted to produce carbon and oxygen in roughly equal quantities by stellar nucleosynthesis, these are the two levels you would have to fix, and your fixing would have to be just about where these levels are actually found to be...A commonsense interpretation of the facts suggests that **a super intellect has monkeyed with physics, as well as chemistry and biology**, and that there are no blind forces worth speaking about in nature. The numbers one calculates from the facts seem to me so overwhelming as to put this conclusion almost beyond question.[38]

Hoyle declared that the inescapable conclusion of this plain truth should not go unnoticed by other scientists.

> I do not believe that any scientist who examined the evidence would fail to draw the inference that **the laws of nuclear physics have been deliberately designed with regard to the consequences they produce inside the stars.**[39]

This plain truth was expressed in the Qur'an 1,400 years ago. Allah indicates the harmony in creation of the heavens in the verse: **Do you not see how Allah created seven heavens in harmony...** (Surah Nuh: 15)

A Lesser Alchemy Lab: The Sun

The conversion of helium into carbon described above is the alchemy of red giants. In smaller stars like our sun, a simpler sort of alchemy takes place. The sun converts hydrogen into helium and this reaction is the source of its energy.

This reaction is no less essential for us to exist than are the reactions in the red giants. Moreover, the sun's nuclear reaction is also a designed process, just like the one in red giants.

Hydrogen, the input element for this reaction, is the simplest element in the universe for its nucleus consists of a single proton. In a helium nucleus, there are two protons and two neutrons. The process taking place in the sun is the fusion of four hydrogen atoms into one helium atom.

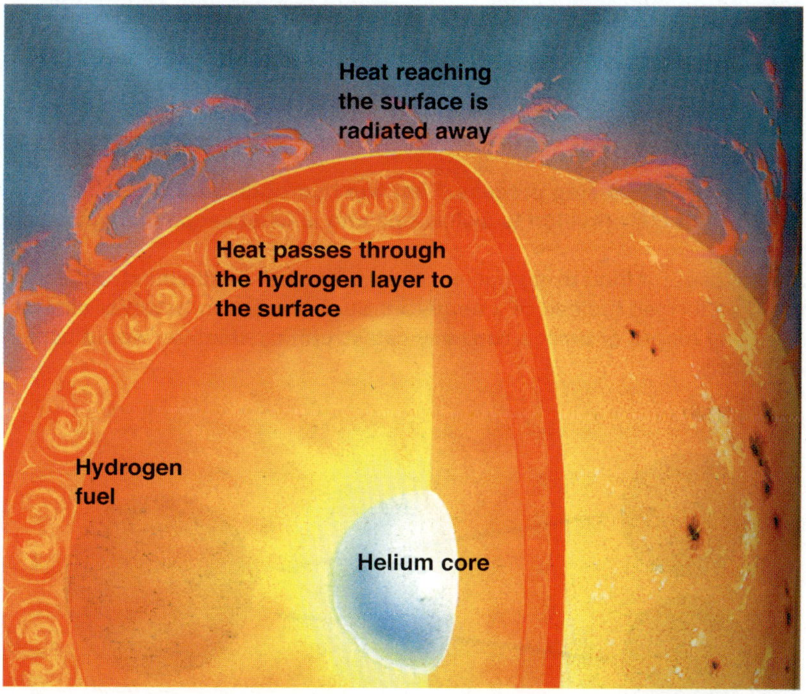

The sun is a giant nuclear reactor that constantly transforms atoms of hydrogen into helium and produces heat in the process. What is crucial to this process however is the incredible precision with which these reactions are balanced within the sun. The slightest change in any of the forces governing these reactions would result in their failure or in a catastrophic runaway explosion.

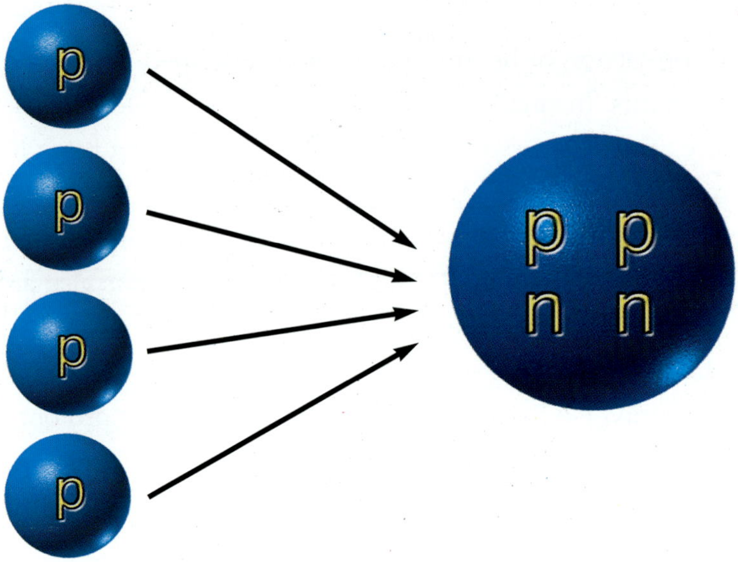

Single-proton hydrogen nuclei

Helium nucleus with two protons and two neutrons

THE CRITICAL REACTION IN THE SUN

1) Above: Four hydrogen atoms in the sun join together to form a single helium atom.

2) Below: This is a two-step process. First two hydrogen atoms fuse forming a deuteron. This transformation is a slow one and is what keeps the sun burning constantly.

3) Opposite page: If the strong nuclear force were just a little bit stronger, a di-proton would be formed instead of a deuteron. Such a reaction however cannot be sustained for any length of time: a runaway catastrophic explosion would occur in just a few seconds.

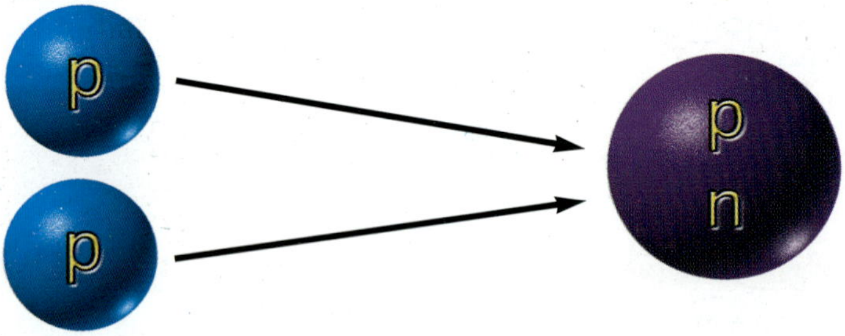

Single-proton hydrogen nuclei

Deuteron nucleus with one proton and one neutron

An enormous amount of energy is released during this process. Nearly all the thermal and light energy reaching Earth is the result of this solar nuclear reaction.

Like the reactions taking place in red giants, this solar nuclear reaction turns out to involve a number of unexpected aspects without which it could not take place. You can't simply jam four hydrogen atoms together and turn them into helium. To make this happen, a two-step process is required, paralleling the one taking place in red giants. In the first step, two hydrogen atoms combine to form an **intermediary nucleus** called **deuteron** consisting of one proton and one neutron.

What force could be great enough to produce a deuteron by jamming two nuclei together? This force is the "strong nuclear force", one of the four fundamental forces of the universe mentioned in the previous section. This is the most powerful physical force in the universe and is billions of billions of billions of billions times stronger than the gravitational force. Nothing but this force could unite two nuclei like this.

Now the really curious thing about all this is that research shows that, strong as it is, the strong nuclear force is just barely strong enough to do what it does. If it were even slightly weaker than it is, it would not be able to unite the two nuclei. Instead, two protons nearing each other would repel each other immediately and the reaction in the sun fizzle out before it ever began. In other words, the sun would not exist as an energy-radiating star. Concerning this, George Greenstein says: **"Had the**

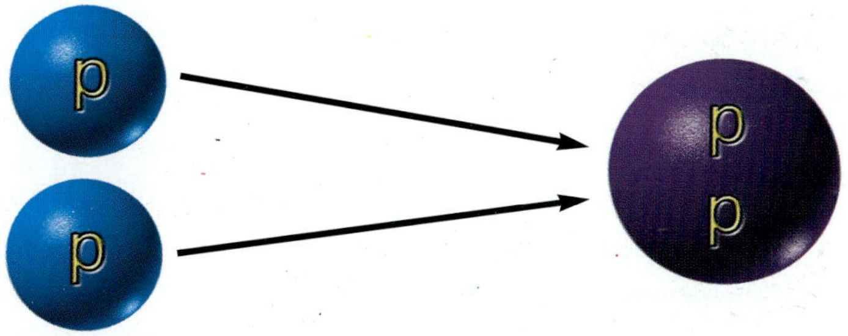

Single-proton hydrogen nuclei

Di-proton nucleus with two protons

strong force had been only slightly less strong, the light of the world would have never been lit."[40]

What, on the other hand, if the strong nuclear force were stronger? To answer that, we first have to look at the process of converting two hydrogen atoms into a deuteron in a little more detail. First, one of the protons is stripped of its electrical charge and becomes a neutron. This neutron forms a deuteron by uniting with a proton. The force causing this unification is the "strong nuclear force"; the force that converts a proton into a neutron on the other hand is a different one and is called the "weak nuclear force". It is weak only by comparison however and it takes about ten minutes to make the conversion. At the atomic level, this is an immensely long time and it has the effect of slowing down the rate at which the reaction in the sun takes place.

Let us now return to our question: What would happen if the strong nuclear force were stronger? The answer is that the reaction in the sun would be changed dramatically because the weak nuclear force would be eliminated from the reaction.

If the strong nuclear force were any stronger than it is, it would be able to fuse two protons to one another immediately and without having to wait ten minutes for a proton to be converted into a neutron. As a result of this reaction, there would be one nucleus with two protons instead of a deuteron. Scientists call such a nucleus a "di-proton". It is a theoretical particle however insofar as it has never been observed to occur naturally. But if the strong nuclear force were much stronger than it is, then there would be real di-protons in the sun. So what? Well by getting rid of the proton-to-neutron conversion, we would be eliminating the "throttle" that keeps the sun's "engine" running as slowly as it does. George Greenstein explains what the result of that would be:

> The Sun would change because the first stage in the formation of helium would no longer be the formation of the deuteron. It would be the formation of the di-proton. And this reaction would not involve the transformation of a proton into a neutron at all. The role of the weak force would be eliminated, and only the strong force would be involved...and as a result the Sun's fuel would suddenly become very good indeed. It would become

so powerful, so ferociously reactive, that the Sun and every other star like it would instantaneously explode.[41]

The explosion of the sun would cause the world and everything on it to burst into flames, burning our blue planet to a crisp in a few seconds. Because the strong nuclear force is precisely fine-tuned to be neither too strong nor too weak, the sun's nuclear reaction is slowed down and the star has been able to radiate light and energy for billions of years. This precise tuning is what makes it possible for mankind to live. If there were even the slightest deviation in this arrangement, the stars (including our sun) would not exist or if they did, they would explode in a short time.

In other words the structure of the sun is neither accidental nor unintentional. Quite the contrary: Allah has created the sun for people to live, as expressed in the verse:

The sun and the moon follow courses (exactly) computed. (Surat ar-Rahman: 5)

Protons and Electrons

So far we have been examining matters concerned with forces that affect atomic nuclei. There is another important equilibrium in the atom that we must consider: the balance between its nucleus and electrons.

Put in its simplest terms, electrons revolve around the nucleus. The reason for this is electrical charge. Electrons have a negative charge and protons have a positive charge. Opposite charges attract, so an atom's electrons are drawn towards the nucleus. But the electrons are also moving at an enormous speed which would, under normal conditions, cause them to shoot away from the nucleus. These two forces (attraction and motion away) are balanced so that the electrons move in orbits around the nucleus.

Atoms are also balanced in terms of their electric charges: the number of orbiting electrons is the same as the number of protons in the nucleus. (For example, oxygen has eight protons and eight electrons.) In this way the electrical force of an atom is balanced and the atom is electrically neutral.

So far, so much basic chemistry. However there is a point in this seemingly simple structure that is overlooked by many. A proton is much bigger than an electron in terms of both size and weight. If an electron were the size of a walnut, a proton would be about the size of a man. Physically, they are quite dissimilar.

But their electrical charges are the same size!

Although their electrical charges are opposite (electrons negative, protons positive) they are also equal. There is no obvious reason why this should be so. Conceivably (and "logically") an electron ought to carry a much smaller charge because it is so much smaller.

But if that were true, then what would happen?

What would happen is that every atom in the universe would be positively charged instead of being electrically neutral. And because like charges repel, every atom in the universe would try and repel every other atom. Matter as we know it could not exist.

What would happen if it suddenly became true now? What would happen if every atom were to start repelling every other?

Quite extraordinary things would happen. Let us begin with the changes that would occur in your body. The moment this change occurred, your hands and your arms holding this book would shatter at once. And not just your hands and arms but also your body, your legs, your eyes, your teeth–every part of your body would explode in a split second.

The room you sit in and the world around you would explode in a moment. All the seas, mountains, the planets in the solar system, and all the stars and galaxies in the universe would shatter into atomic dust. And there would never again be anything in the universe to observe. The universe would become a mass of disorganized atoms pushing each other around.

By how much would the sizes of the electrical charges of protons and electrons have to differ in order for this dreadful thing to happen? One percent? A tenth of one percent? George Greenstein addresses this question in *The Symbiotic Universe*:

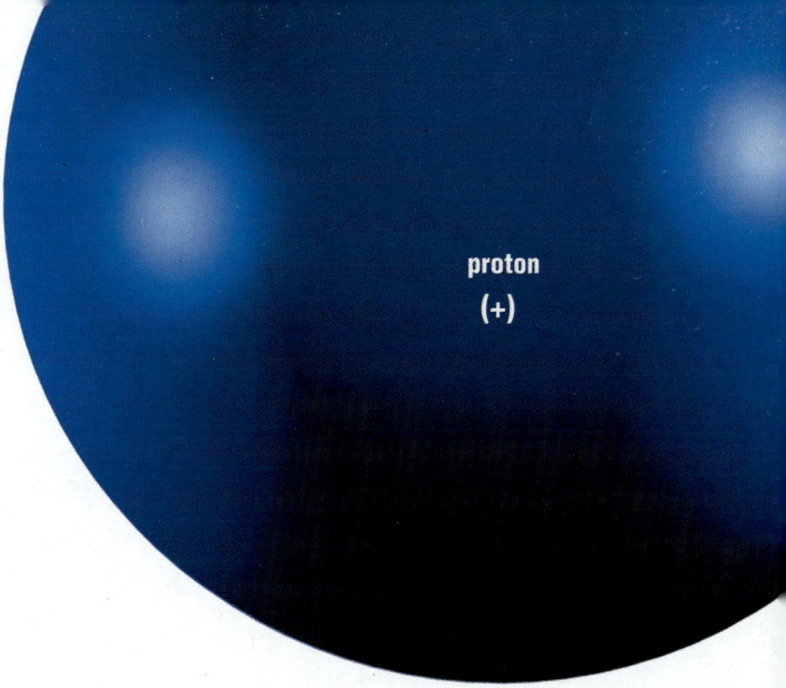

Both the mass and the volume of a proton are incomparably larger than those of an electron but, strangely enough, these two particles have equal (though opposite) electrical charges. Because of this fact, atoms are electrically neutral.

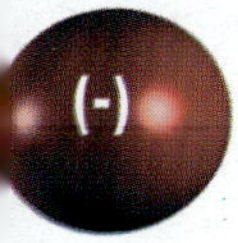

> Small things like stones, people, and the like would fly apart if the two charges differed by as little as one part in 100 billion. Larger structures like the Earth and the Sun require for their existence a yet more perfect balance of one part in a billion billion.[42]

Here is yet another precisely-tuned equilibrium that proves that the universe is intentionally designed and created for a particular purpose. As John D. Barrow and Frank J. Tipler maintain in their book *"The Anthropic Cosmological Principle"*, **"there is a grand design in the Universe that favours the development of intelligent life."**[43]

Of course every design proves the existence of a conscious "designer". That is Allah alone, "Lord of all the worlds", described in the Qur'an as the only Power Who created the universe from nothingness, and designed and fashioned it as He willed. As stated in the Qur'an, **"He built the heaven, He raised its vault high and made it level." (Surat an-Nazi'at: 27-28)**

Thanks to the extraordinary balances that we have seen in this chapter, matter is able to remain stable and this stability is evidence of the perfection of Allah's creation as revealed in the Qur'an:

> **Everyone in the heavens and the earth belongs to Him. All are submissive to Him. (Surat ar-Rum: 25)**

CHAPTER IV

THE ORDER IN THE SKIES

...Something else has to be behind things, somehow guiding them. And that, one might say, is a kind of mathematical proof of divinity.

Guy Marchie, American Science Writer[44]

During the night of July 4th in 1054, Chinese astronomers witnessed an extraordinary event: a very bright star that suddenly appeared near the constellation Taurus. It was so bright that it could easily be seen even in daytime. At night it was brighter than the moon.

What Chinese astronomers observed was one of the most interesting and catastrophic astronomic phenomena in our universe. It was a supernova.

A supernova is a star that is shattered by an explosion. A huge star destroys itself in an immense blast and the material of its core is scattered in every direction. The light produced during this event is a thousand times brighter than normal.

Scientists today think that supernovas play a key role in the formation of the universe. These explosions are what cause different elements to be carried to different parts of the universe. It is supposed that the material ejected by these explosions subsequently combines to form a new galaxy or a star somewhere else in the universe. According to this hypothesis, our solar system, the sun and its planets including Earth, are the products of some incredibly ancient supernova.

Although supernovas may seem to be ordinary explosions, they in fact are minutely structured in their details. In *Nature's Destiny* Michael Denton writes :

> The distances between supernovae and indeed between all stars is critical for other reasons. The distance between stars in our galaxy is about 30 million miles. If this distance was much less, planetary orbits would be destabilized. If it was much more, then the debris thrown out by a supernova would be so diffusely distributed that planetary systems like our own would in all probability never form. If the cosmos is to be a home for life, then the flickering of the supernovea must occur at a very precise rate and the average distance between them, and indeed between all stars, must be very close to the actual observed figure.[45]

The ratio of supernovas and stars' distances are just two more of the fine-tuned details of this miraculous universe. Examining deeper the uni-

The gigantic explosions known as supernova cause matter to move throughout the universe. The enormous distances between the universe's stars and galaxies moderate the risk that such an explosion will affect other bodies.

verse the arrangement we see is beautiful both in the organization and design.

Why is There So Much Space?

Let's recap a few points that we made earlier. The universe following the Big Bang was a nebula of just hydrogen and helium. Heavier elements were produced later by means of intentionally-designed nuclear reactions. Yet the existence of heavier elements is not a sufficient reason for the universe to become a suitable place for life. A much more important issue is how the universe was formed and ordered.

We shall start by asking how big the universe is.

We have adorned the nearest
heaven with an adornment,
the stars.
(Surat as-Saffat: 6)

The planet Earth is a part of the solar system. In this system there are nine major planets with fifty-four satellites, and an uncounted number of asteroids all revolving around a single star called "Sun", a middle-sized star compared with others in the universe. Earth is the third planet from the sun.

Let us first try to understand the size of this system. The diameter of the sun is 103 times that of the earth. To visualize this, the planet Earth has diameter of 12,200 kms. If we scaled that down to the dimensions of a glass ball, the sun would be about the size of soccer ball. But the interesting thing is the distance between the two. Keeping to the same scale, the two balls should be 280 meters apart. Some of the objects representing the outer planets would have to be set several kilometers away.

Big though this might seem, the solar system is a quite miniscule in size compared with the Milky Way, the galaxy in which it is located. There are over 250 billion stars in the Milky Way–some similar to the sun, others bigger, others smaller. The star nearest to the sun is Alpha Centauri. If we wanted to add Alpha Centauri in our model system, it would have to be located 78,000 kilometers away.

That's too big for almost anyone to grasp, so let's reduce the scale. We'll assume the earth to be as big as a dust-particle. That would make the sun as big as a walnut about three meters from the earth. On this scale, Alpha Centauri would have to be located 640 kilometers from the sun.

The Milky Way consists of about 250 billion stars with similarly mind-boggling distances between them. The sun is located closer to the edge of this spiral-shaped galaxy than it is to the center.

Even the Milky Way is dwarfed by the vast size of the whole universe. It is just one of many galaxies–nearly 300 billion of them according to recent calculations. And the distances between galaxies are millions of times greater than that between the sun and Alpha Centauri.

George Greenstein, in The Symbiotic Universe, comments on this unimaginable vastness:

> Had the stars been somewhat closer, astrophysics would not have been so very different. The fundamental physical processes occurring within stars,

> nebulas, and the like would have proceeded unchanged. The appearance of our galaxy as seen from some far-distant vantage point would have been the same. About the only difference would have been the view of the night time sky from the grass on which I lie, which would have been yet richer with stars. And oh, yes-one more small change: There would have been no me to do the viewing...All that waster space! On the other hand, in this very waste lies our safety.[46]

Greenstein also explains the reason for this. In his view, the huge distances in space makes it possible for certain physical variables to be arranged so as to be exactly suitable for human life. He also notes the importance of this huge space in allowing Earth to exist while minimizing the risk of collision with other stars.

In short, the distribution of celestial bodies in space is exactly what it must be for human life to exist on our planet. These huge spaces are the outcome of an intentional design for a purpose and not a result of coincidence.

Entropy and Order

In order to understand the concept of order in the universe, we need first to talk about the Second Law of Thermodynamics, one of the fundamental universal physical laws.

This law states that, left to themselves, organized systems will become unstable and less organized as time advances. This law is also called the Law of Entropy. In physics, entropy is the amount of disorder in a system. The transition of a system from a stable condition into an unstable condition is the same as an increase in its entropy. The instability is directly related to the entropy of that system.

This is commonplace knowledge, many examples of which we may observe in our daily lives. If you abandon a car in some exposed place for a year or even a couple of months, you certainly wouldn't expect it be in just as good condition as you left it when you return. You'll probably notice flat tires, broken windows, corroded parts in the engine and body, etc. Similarly if you neglect to straighten up your house for a few days and

An abandoned car deteriorates and falls apart. Everything in the universe is subject to entropy: the law says that, left to itself, everything becomes less stable and less organized with the passage of time.

you'll immediately see it getting dustier and more disorganized as time goes by. This is a kind of entropy; however you can undo it by cleaning and picking things up and by taking out the trash.

The Second Law of Thermodynamics is widely accepted as valid and binding. Einstein, the most important scientist of our century, said that this law was the "first law of all sciences". The American scientist Jeremy Rifkin comments in *Entropy: A New World View:*

> The Entropy Law will preside as the ruling paradigm over the next period of history. Albert Einstein said that **it is the premier law of all science**: Sir Arthur Eddington **referred to it as the supreme metaphysical law of the entire universe**.[47]

It is important to note that the Law of Entropy by itself renders many of the claims of materialism invalid right from the start. For if there is a definite design and order in the universe, the law holds that, in the course of time, this situation will be undone by the universe itself. There are two conclusions to be reached from this observation:

1) Left to itself, the universe cannot exist for eternity. The second law says that without external intervention of some sort, entropy will eventually be maximized throughout the universe causing it to assume a completely homogenous state.

2) The claim that the order we observe is not the result of external intervention is also invalid. Immediately after the Big Bang, the universe was in precisely such a completely disorganized state as would exist if entropy had been maximized. But that has changed as we can plainly see by looking around. That change took place in violation of one of nature's fun-

Every galaxy in the universe is proof of the organized structure that everywhere exists. These magnificent systems, with an average of 300 billion stars each, display an evident balance and harmony.

damental laws–the Law of Entropy. There is simply no way to account for this change except to posit some sort of supernatural creation.

An example will perhaps make the second point clearer. Imagine the universe to be a huge cave full of a jumble of water, rocks, and dirt. We leave the cave alone for several billion years and then come back and take a look at it. Upon our return we notice that some of the rocks have gotten smaller, some have disappeared, the level of dirt is higher, there's more

NOBEL PRIZE WINNER PHYSICIST MAX PLANCK: " A certain order prevails in our universe. This order can be formulated in terms of purposeful activity"

mud, and so on. Things are more disordered, which is normal–just as we might expect. If, billions of years later, you find rocks delicately carved into statues, you would definitely decide that this order cannot be explained away by laws of nature. The only rational explanation is that "a conscious mind" caused these things to be.

So the order of this universe is the most overwhelming proof of the existence of a superior consciousness. The Nobel prize winner German physicist Max Planck explains the order in the universe:

> At all events, we should say, in summing up, that, according to everything taught by the exact sciences about the immense realm of nature in which our tiny planet plays an insignificant role, a certain order prevails - one independent of the human mind. Yet, in so far as we are able to ascertain through our senses, **this order can be formulated in terms of purposeful activity**. There is evidence of an intelligent order of the universe.[48]

Paul Davies explains the triumph of this marvelous equilibrium and harmony over materialism thus:

> Everywhere we look in the Universe, from the far flung galaxies to the deepest recesses of the atom, we encounter order... Central to the idea of a very special, orderly Universe is the concept of *information*. A highly structured system, displaying a great deal of organised activity, needs a lot of information to describe it. Alternatively, we may say that it contains much information.
>
> We are therefore presented with a curious question. If information and order always has a natural tendency to disappear, **where did all the informa-**

tion that makes the world such a special place come from originally? The Universe is like a clock slowly running down. **How did it get wound up in the first place?**[49]

Einstein refers to this order as an unexpected event, and also says that it should be regarded as a miracle:

> Well, a priori [reasoning from cause to effect] one should expect that the world would be rendered lawful [obedient to law and order] only to the extent that we [human beings] intervene with our ordering intelligence... [But instead we find] in the objective world a high degree of order that we were a priori in no way authorized to expect. This is the 'miracle' that is strengthened more and more with the development of our knowledge.[50]

In short, the order in the universe demands deep and extensive understanding and knowledge. It is designed, organized, and preserved by Allah.

Allah reveals how the heavens and earth are preserved by His supreme power in the Qur'an:

> **Allah keeps a firm hold on the heavens and earth, preventing them from vanishing away. And if they vanished no one could then keep hold of them. Certainly He is Most Forbearing, Ever -Forgiving. (Surah Fatır: 41)**

The divine order in this universe reveals the weakness of the materialistic belief of a universe that is a mass of ungoverned matter. This is revealed in another verse:

> **If the truth were to follow their whims and desires, the heavens and the earth and everyone in them would have been brought to ruin...(Surat al-Muminun: 71)**

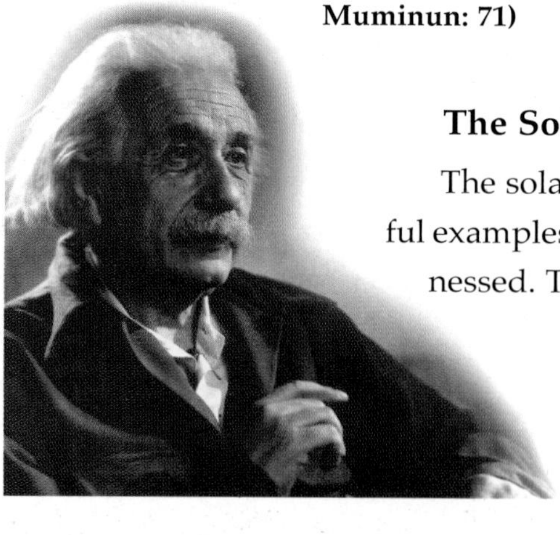

The Solar System

The solar system is one of the most wonderful examples of this beautiful harmony to be witnessed. There are nine planets with fifty-four

ALBERT EINSTEIN:
"We find in the objective world a high degree of order."

known satellites and an unknown number of smaller bodies. The major planets counting outward from the sun are Mercury, Venus, Earth, Mars, Jupiter, Saturn, Uranus, Neptune, and Pluto. Earth is the only one on which life is known to exist. It is surely the only one on which human beings can live and survive unaided thanks to abundant land and water and to a breathable atmosphere.

In the structure of the solar system, we encounter another beautiful example of equilibrium: the balance between a planet's **centrifugal force** countered by the **gravitational attraction** of its primary. (In astronomy, a primary is something that another body revolves about. Earth's primary is the sun; the moon's primary is Earth.) Without this balance, everything in the solar system would fly off into the chilling depths of outer space. The balance between these two forces results in paths (orbits) that the planets and other bodies follow around their primaries. If a body moved at too slow a speed, it would plunge into the primary; if it moved at too fast a speed, the primary would be unable to hold onto it, and it would fly off into space. Instead, every body moves at just the right speed to keep it in orbit. Moreover, this equilibrium has to be different for each body because the distance of planets to the sun differs. So do their masses. Therefore, they have to have different orbital speeds not to plunge into the sun or not to fly off into space.

Materialist astronomy holds that the origin and survival of the solar system can be explained by coincidence. Over the last three centuries, many of its adherents have speculated on how this marvelous order should have come to pass and they have failed to get anywhere. To a materialist, the equilibrium and order of the solar system are inexplicable mysteries.

Isaac Newton, one of the pioneers and founders of modern physics and astronomy, saw in the structure of the universe magnificent evidence of divine creation.

THE SOLAR SYSTEM

It is not for the sun to catch up the moon nor for the night to outstrip the day. Each (just) swims along in (its own) orbit. (Surah Ya Sin: 40)

Astronomers like Kepler and Galileo, among the first to discover this superlative equilibrium, acknowledged it as a deliberate design and a sign of divine intervention in the whole universe. Isaac Newton, recognized as one of the greatest scientific minds of all times, once wrote:

> This most elegant system of suns, planets, and comets could arise from the purpose and sovereignty of an intelligent and mighty being...He rules them all, not as a soul but as a sovereign lord of all things, and because of His sovereignty He is commonly called "Lord God Almighty."[51]

The Place of the Earth

Besides this wonderful equilibrium, the place of earth in the solar system and in the universe is also another piece of evidence of a perfect act of creation on Allah's part.

The latest astronomical findings have shown the importance of the other planets' existence for Earth. Jupiter's size and position turn for example out to be critical. Astrophysical calculations show that, as the biggest planet in the system, Jupiter supplies stability to the orbits of Earth and all the other planets. Jupiter's protective role over the earth is explained in an article "How special Jupiter is" by George Wetherill:

> Without a large planet positioned precisely where Jupiter is, the earth would have been struck a thousand times more frequently in the past by comets and meteors and other interplanetary debris. If it were not for Jupiter, we wouldn't be around to study the origin of the solar system.[52]

To put it briefly, the structure of the solar system was specially designed for mankind to live.

Let us also consider the place of solar system in the universe. Our solar system is located in one of the huge spiral arms of the Milky Way, closer to the edge than to the center. What advantage could there be in that? *In Nature's Destiny*, Michael Denton explains:

> What is so striking is that the cosmos appears to be not just supremely fit for our own being and for our biological adaptations, but also for our understanding... Because of the position of our solar system on the edge of the galactic rim, we can gaze farther into the night to distant galaxies and gain

knowledge of the overall structure of the cosmos. Were we positioned in the center of a galaxy, we would never look on the beauty of a spiral galaxy nor would we have any idea of the structure of our universe.[53]

In other words, even Earth's location in the galaxy is evidence that it was intended for mankind to live on, no less than are all the other physics laws of the universe.

It is the plain truth that the universe is created and arranged by Allah.

The reason that some people cannot understand this point is their own prejudice. But any objective mind without prejudice will easily understand that the universe is created and organized by Allah for mankind to live in, just as is revealed:

We did not create heaven and earth and everything in between them to no purpose. That is the opinion of those who are disbelivers. Woe to those who are disbelievers, because of the Fire! (Surah Sad:27)

This deep understanding is revealed in yet another verse of the Qur'an:

In the creation of the heavens and the earth, and the alternation of night and day, there are Signs for people with intelligence: those who remember Allah, standing, sitting and lying on their sides, and reflect on the creation of the heavens and the earth: 'Our Lord, You have not created this for nothing. Glory to You!" (Surat Al-'Imran: 190-191)

CHAPTER V

THE BLUE PLANET

The Earth, with its atmosphere and oceans, its complex biosphere, its crust of relatively oxidised, silica rich, sedimentary, igneous, and metamorphic rocks overlaying [a magnesium silicate mantle and core] of metallic iron, with its ice caps, deserts, forests, tundra, jungles, grasslands, fresh-water lakes, coal beds, oil deposits, volcanoes, fumaroles, factories, automobiles, plants, animals, magnetic field, ionosphere, mid-ocean ridges, convincing mantle... is a system of stunning complexity.

J. S. Lewis, American Geologist[54]

An imaginary space-traveler approaching the solar system from interstellar space would encounter a very interesting scene. Let us imagine that we are such travelers and that we're arriving at the plane of the ecliptic–the great circle of the celestial sphere in which all the major planets of our solar system move. The first planet we will meet is Pluto. This planet is quite a cold place. The temperature is around -238°C. The planet has a thin of atmosphere that is in a gaseous state only when it draws slightly nearer to the sun in its rather elliptical orbit. At other times, the atmosphere becomes a mass of ice. Pluto, briefly, is a lifeless sphere enveloped in ice.

Advancing towards the sun, you next encounter Neptune. It is cold too: approximately -218°C. The atmosphere, consisting of hydrogen, helium and methane, is poisonous for life. Winds blowing nearly 2,000 kilometers an hour blast across the surface of the planet.

Next is Uranus: a gaseous planet with rocks and ice on its surface. The temperature is -214°C and the atmosphere again consists of hydrogen, helium and methane--unsuitable for human beings to live in.

You reach Saturn after Uranus. This is the second biggest planet in the solar system and is particularly notable for the system of rings encircling it. These rings are made up of gases, rock and ice. One of the many interesting things about Saturn is that it is composed entirely of gas: 75% hydrogen and 25% helium and its density is less than that of water. If you want to "land" on Saturn, you'd better design your spaceship to be like an inflatable boat! The average temperature is again very low: -178°C.

Coming up next is Jupiter: the biggest planet in the solar system, it is 318 times the size of Earth. Like Saturn, Jupiter is also a gaseous planet. Since it is difficult to distinguish between "atmosphere" and "surface" on such planets, it is hard to say what the "surface temperature" is but in the upper reaches of the atmosphere, the temperature is -143°C. A notable feature of Jupiter's atmosphere is something called the Great Red Spot. It was first noticed 300 hundred years ago. Astronomers now know that it is an enormous storm system that has been raging in the Jovian atmosphere for centuries. It is big enough to swallow up a couple of planets the size of

Allah created the heavens and the earth with truth. There is certainly a Sign in that for the believers. (Surat al-Ankabut: 44)

Earth whole. Jupiter may be a visually thrilling planet, but it's no home for people, who would be killed instantly by its freezing temperatures, violent winds, and intense radiation.

Then comes Mars. The atmosphere of Mars cannot sustain human life because it is mostly carbon dioxide. The surface is everywhere pocked with craters: the result of eons of meteor impacts and strong winds blowing across the surface that can raise sandstorms that last for days or weeks at a time. The temperature varies rather much but drops as low as -53°C. There has been much speculation that Mars might harbor life, but all the evidence shows that this is a lifeless world too.

Speeding away from Mars and heading toward the sun, we notice a blue planet that we decide to skip for the time being while we explore some more. Our search brings us to a planet called Venus. This planet is everywhere shrouded in brilliant white clouds but the temperature at the surface is 450°C, which is enough to cause lead to melt. The atmosphere

is composed mostly of carbon dioxide. At the surface, the atmospheric pressure is equal to 90 terrestrial atmospheres: on Earth, you'd have to descend a kilometer into the sea before you reached a pressure that high. The atmosphere of Venus contains layers of gaseous sulfuric acid several kilometers deep. When it rains on Venus, it isn't raining rain you know: it's raining acid. No human or other life could exist in such a hellish place for a second.

We press on and come to Mercury, a small, rocky world, blasted by the heat and radiation of the sun. Its rotation has been so slowed down by its proximity to the sun that the planet makes only three full axial rotations in the time it takes to revolve twice around the sun. In other words, two of Mercury's "years" is equal to three of its "days". Because of this prolonged diurnal cycle, one side of Mercury becomes extremely hot while the other is extremely cold. The difference between the daytime and nighttime sides of Mercury is as much as 1,000°C. Of course such an environment cannot support life.

Even Mars, the only other planet in the solar system to come close to resembling the earth physically, is nothing but an arid, lifeless ball of rock.

THE INFERNAL SURFACE OF VENUS

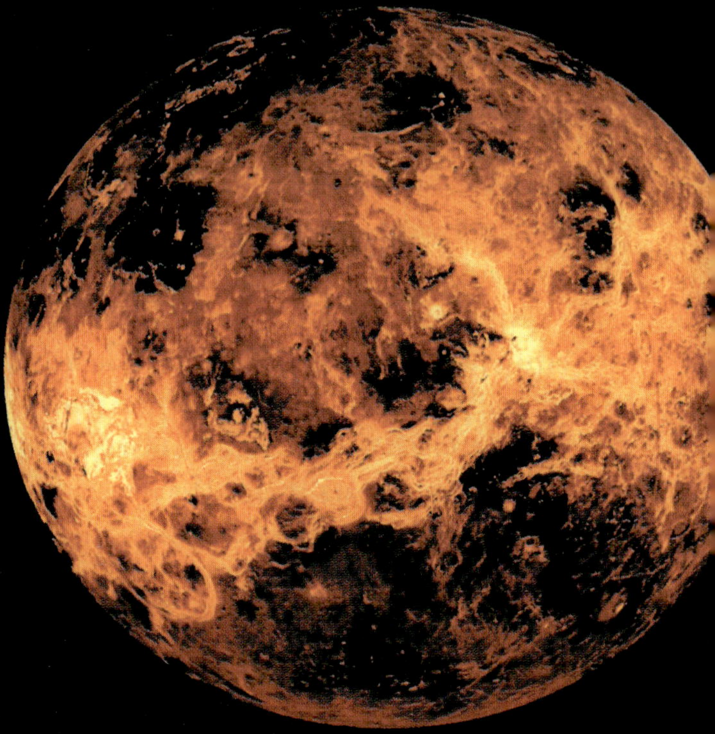

The surface temperature on Venus reaches as high as 450° C, which is sufficient to melt lead. The surface of this world resembles a ball of fire covered with lava. Its atmosphere is thick with sulfuric acid and a sulfuric acid rain falls constantly. The atmospheric pressure at the surface is 90 times that of Earth: the equivalent of a depth of 1,000 meters beneath the sea.

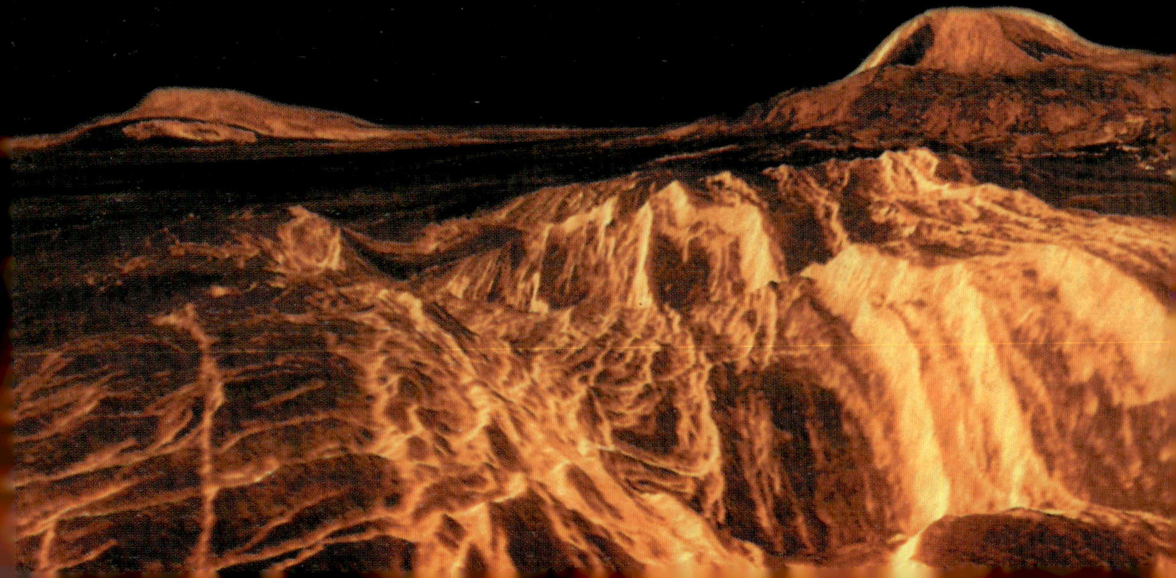

To sum up, we've taken looks at eight planets and not one of them, including their fifty-three satellites offers anything that might serve as a haven for life. Each of them is lifeless ball of gas, ice, or rock.

But the blue planet that we skipped over a while ago? That one's very different from the others. With its hospitable atmosphere, surface features, ambient temperatures, magnetic field, and supply of elements and set just the right distance from the sun, it almost seems as if it had been specially created to be a home for life.

And, as we shall discover, it was.

A Brief Digression and Warning About "Adaptation"

In the rest of this chapter we will be examining features of Earth that make it clear that our planet was created specifically for the support of life. But before we do that, we need to make a brief digression in order to avoid the possibility of any misunderstanding. This digression is especially for those who are in the habit of recognizing the theory of evolution as a scientific truth and who strongly believe in the concept of "adaptation".

"Adaptation" is the noun form of the verb "adapt". "Adapt" implies a modification according to changing circumstances. As used by evolutionists, it means a "modification of an organism or its parts that makes it more fit for existence under the conditions of its environment". The theory of evolution claims that all life on earth is derived from a single organism (a single common ancestor) that itself came into being as a result of chance and the theory makes heavy use of this sense of the word "adaptation" to support its case. Evolutionists hold that living organisms change into new species by adapting to their environment. We have discussed the invalidity of this claim, that mechanisms of adaptation to natural conditions in living beings come into play only under certain circumstances and it can never transform one species into another in detail in our other books.[55] (This is summed up in the appendix "*Evolution Deceit*" in this book) The theory of evolution with its concept of "adaptation" is really just a form of Lamarckism, a theory of organic evolution that holds that environmental changes cause structural changes in animals and plants that can be transmitted to offspring- a theory that has been soundly and rightly dismissed by scientific circles.

Yet even though it has no scientific basis, the idea of adaptation impresses most people and that is why we must address this point here before going on. From belief in the adaptability of life-forms, it is only a step to the idea that life could have developed on other planets as well as it did once on Earth. The possibility of little green creatures living on Pluto who might work up a slight sweat when the temperature soared to 238°C, who breathe helium instead of oxygen, and who drink sulfuric acid instead of water somehow tickles people's fancy, especially people whose fancies have been richly nourished by the products of Hollywood studios.

But these are only such stuff as dreams (and Hollywood movies) are made of however and evolutionists who are better informed about biology and biochemistry do not even attempt to defend such notions. They know quite well that life exists only if necessary conditions and elements are available. If they really believe in them at all, the partisans of the little green men (or other alien life-forms) are those who blindly adhere to the theory of evolution and are ignorant of even the basics of biology and biochemistry and who, in their ignorance, come up with preposterous scenarios.

So in understanding the error in the concept of adaptation, the first thing that we need to note is that **life can only exist if certain essential conditions and elements are present**. The only model of life that is based on scientific criteria is that of **carbon-based life** and scientists are in agreement that there is no other form of life to be found anywhere elsewhere in the universe.

Carbon is the sixth element in the periodic table. This atom is the basis of life on earth because all organic molecules (such as nucleic acids, amino acids, proteins, fats, and sugars) are formed by the combination of carbon with other elements in various ways. Carbon forms millions of different types of proteins by combining with hydrogen, oxygen, and nitrogen etc. No other elements can take the place of carbon. As we shall see in the sections ahead, no element but carbon has the ability to form the many different kinds of chemical bonds on which life depends.

Consequently if life is going to exist on any planet anywhere in the universe it is going to have to be carbon-based.[56]

There are a number of conditions that are absolutely essential in order for carbon-based life to exist. For example, carbon-based organic compounds (like proteins) can exist only within a certain range of temperatures. They start to dissociate over 120°C and are irrecoverably damaged if they are frozen below -20°C. But it is not only temperature that plays a vital role in determining the allowable limits of suitable conditions for carbon-based life to exist: so too do the type and amount of light, the strength of gravity, the composition of the atmosphere, and the strength of the magnetic field. Earth provides precisely such conditions as are needed to make life possible. If even one of conditions were to be changed, if average temperatures surpassed 120°C for example, there would be no life on Earth.

Therefore our little green creatures who might work up a slight sweat when the temperature soars to 238°C, who breathe helium instead of oxygen, and who drink sulfuric acid instead of water are not going to exist anywhere because carbon-based life-forms cannot survive under such conditions and carbon-based life-forms are the only kind there is. Life can only exist in an environment within limits and under conditions that are deliberately designed for life. That is true of life in general and of human beings in particular.

Earth is such a deliberately-designed environment.

The Temperature of the World

Temperature and atmosphere are the first essential factors for life on Earth. The Blue Planet has both a temperature that is livable and an atmosphere that is breathable for living things, especially for such complex living things as human beings. These two extremely different factors however have come into being as a result of conditions that turn out to be ideal for both.

Unlike the other 63 major planets and satellites in our solar system, the planet Earth is the only one possessing an atmosphere, an ambient temperature, and a surface suitable for life. Although liquid water, a fundamental requirement for life, is found nowhere else in the solar system, three-fourths of the earth's surface is covered with it.

One of these is the distance between the earth and the sun. Earth could not be a home for life if were as near the sun as Venus is or as far from it as Jupiter: carbon-based molecules can only survive between the limits of 120 and –20°C and Earth is the only planet whose average temperatures fall within those limits.

When one considers the universe as a whole, coming across a range of temperatures as narrow as this is quite a difficult task because temperatures in the universe vary from the millions of degrees of the hottest stars to absolute zero (-273°C). In such a vast range of temperatures, the thermal interval that allows life to exist is slim indeed; but the planet Earth has it.

The American geologists Frank Press and Raymond Siever draw attention to the average temperatures prevailing on Earth. They note that "**life as we know it is possible over a very narrow temperature interval.** This interval is perhaps 1 or 2 percent of the range between a temperature of absolute zero and the surface temperature of the Sun." [57]

The maintenance of this thermal range is also related to the amount of heat that the sun radiates as well as to the distance between the earth and the sun. According to calculations, a reduction of just 10% in the sun's radiant energy would result in the earth surface's being covered by layers of ice many meters thick and that if it were to increase by a little, all living things would be scorched and die.

Not only must the average temperature be ideal: the available heat must also be distributed fairly equally over the whole planet. A number of special precautions have been taken to ensure that this in fact happens.

The earth's axis is inclined 23° 27'to the plane of the ecliptic. This inclination prevents overheating of the atmosphere in the regions between the poles and the equator, causing them to become more temperate. If this inclination did not exist, the temperature gradient between the poles and equator would be much higher than it is and the temperate zones wouldn't be so temperate–or livable.

The rotational speed of Earth on its axes also helps keep the thermal distribution in balance. The earth makes a complete rotation once every

Many completely different factors such as the distance between Earth and Sun, the planet's rotational speed, the inclination of its axes, and the geographical features of the surface all combine to ensure that our world is heated in just the right way that life needs and that this heat is adequately distributed.

24 hours with the result that alternating periods of daylight and darkness are fairly short. Because they are short, the thermal gradient between the light and dark sides of the planet are quite modest. The importance of this can be seen in the extreme example of Mercury, where a day lasts longer than a year and where the difference between daytime and nighttime temperatures is almost 1,000°C.

Geography also helps distribute heat equally over the earth. There is a difference of about 100°C between the polar and equatorial regions of Earth. If such a thermal gradient were to exist over a completely level area, the result would be winds reaching speeds as high as 1,000 kilometers an hour sweeping away everything in their path. Instead, Earth is full of geographical barriers that block the huge movements of air that such a thermal gradient would otherwise cause. Those barriers are chains of mountains like the one that stretches from the Pacific in the east to the Atlantic in the west, beginning with the Himalayas in China and continuing with the Taurus mountains in Anatolia and the Alps in Europe. At sea, the excess heat in the equatorial regions is transferred north and south thanks to the superior ability of the water to conduct and dissipate heat.

At the same time, there are a number of auto-control systems that help keep the atmospheric temperature in balance. For example when a region heats up, the rate at which its water vaporizes increases, causing clouds to form. These clouds reflect more light back into space, preventing both the air and the surface below from getting warmer.

The Mass of the Earth and the Planet's Magnetic Field

The size of Earth is no less important for life than are its distance from the Sun, its rotational speed, or geographical features. Looking at the planets we see a great range of sizes: Mercury is less than a tenth the size of Earth while Jupiter is 318 times bigger. Is the size of Earth as compared with other planets "coincidental"? Or is it deliberate?

When we examine the dimensions of Earth we can easily see that our planet was designed to be exactly as big as it is. American geologists Frank Press and Raymond Siever comment on Earth's "fitness":

> **And Earth's size was just about right** -not too small as to lose its atmosphere because its gravity was too small to prevent gasses from escaping into space, and not so large that its gravity would hold on to too much atmosphere, including harmful gases.[58]

In addition to its mass, the interior of Earth is also specially designed. Because of its core, Earth has a strong magnetic field whose role in the preservation of life is vital. According to Press and Siever:

> The earth's interior is a gigantic but delicately balanced heat engine fueled by radioactivity ...Were it running more slowly, geological activity would have proceeded at a slower pace. Iron might not have melted and sunk to

> form the liquid core, and the magnetic field would never have developed...if there had been more radioactive fuel and a faster running engine, volcanic gas and dust would have blotted out the Sun, the atmosphere would have been oppressively dense, and the surface would have been racked by daily earthquakes and volcanic explosions.[59]

The magnetic field these geologists talk about is of great importance for life. This magnetic field originates from the structure of Earth's core. The core consists of heavy elements like iron and nickel that are capable of holding a magnetic charge. The inner core is solid while the outer one is liquid. The two layers of the core move around each other and this move-

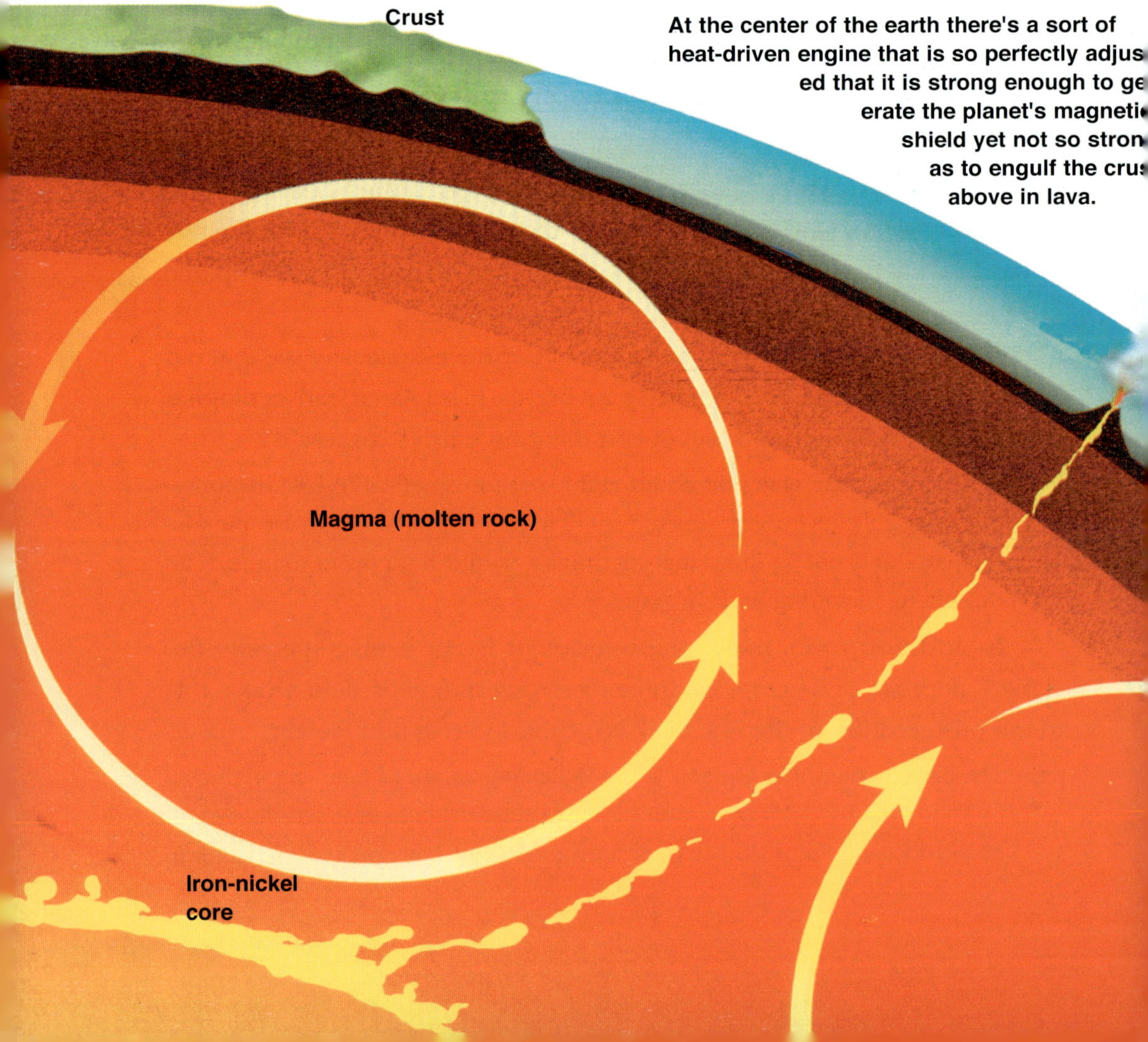

ment is what generates Earth's magnetic field. Extending far beyond the surface, this field protects Earth from the effects of detrimental radiation from outer space. The radiation of stars other than the sun cannot travel through this shield. The **Van Allen Belt**, whose magnetic lines extend ten thousand miles from Earth, protects the globe from this deadly energy.

It is calculated that the plasma clouds trapped by the Van Allen Belt sometimes attain energy levels 100 billion times more powerful than that the atomic bomb released over Hiroshima. Cosmic rays may be equally detrimental. The earth's magnetic field however lets only 0.1% of that radiation through and that is absorbed by the atmosphere. The electrical energy needed to create and maintain such a magnetic field is nearly a billion amperes, as much as mankind has generated throughout history.

If this protective shield did not exist, life would be destroyed by harmful radiation from time to time and might not have come into existence at all. But as Press and Siever point out, Earth's core is exactly designed to keep the planet safe.

In other words, there is a special purpose as stated in the Qur'an:

We made the sky a persevered and protected roof yet still they turn away from Our Signs. (Surat al-Anbiya: 32)

The Fitness of the Atmosphere

As we have seen, Earth's physical features--mass, structure, temperature and so on–are "just right for life". Such features alone are not enough to allow life to exist on Earth however. Another vital factor is the composition of the atmosphere.

We noted above how science-fiction movies sometimes mislead people. One example of how they do this is how easily space travelers and explorers come across planets with breathable atmospheres: they seem to be lying all over the place. If we could explore the real universe, we'd discover that this isn't true at all: the possibility of another planet's having an atmosphere that we could breathe is most unlikely. That's because the atmosphere of Earth is specially designed to support life in a number of crucial ways.

Earth's atmosphere as seen from above by NASA astronauts while over the Philippines.

The atmosphere of Earth is composed of 77% nitrogen, 21% oxygen, and 1% carbon dioxide. Let's start with the most important gas: oxygen. Oxygen is vitally important to life because it enters into most of the chemical reactions that release the energy that all complex life-forms require.

Carbon compounds react with oxygen. As a result of these reactions, water, carbon dioxide, and energy are produced. Small "bundles" of energy that are called ATP (adenosine triphosphate) and are used in living cells are generated by these reactions. This is why we constantly need oxygen to live and why we breathe to satisfy that need.

The interesting aspect of this business is that the percentage of oxygen in the air we breathe is very precisely determined. Michael Denton writes on this point:

> Could your atmosphere contain more oxygen and still support life? No! Oxygen is a very reactive element. Even the current percentage of oxygen in the atmosphere, 21 percent, is close to the upper limit of safety for life at ambient temperatures. The probability of a forest fire being ignited increases by as much as 70 percent for every 1 percent increase in the percentage of oxygen in the atmosphere.[60]

According to the British biochemist James Lovelock:

> Above 25% very little of our present land vegetation could survive the raging conflagrations which would destroy tropical rain forests and arctic tundra alike... The present **oxygen level is at a point where risk and benefit nicely balance.**[61]

That the proportion of oxygen in the atmosphere remains at this precise value is the result of a marvelous "recycling" system: Animals constantly consume oxygen and produce carbon dioxide, which, for them, is not breathable. Plants do just the opposite: they take in carbon dioxide, which they need to live, and release oxygen instead. Thanks to this system, life goes on. Plants release millions of tons of oxygen into the atmosphere every day.

Without the cooperation and balance of these two different groups of living things, our planet would be unlivable. For example, if living things only took in carbon dioxide and released oxygen, the earth's atmosphere

Even a 5% increase in the amount of oxygen in our planet's atmosphere would result in fires that would destroy much of its forests.

would support combustion much more easily than it does and even a tiny spark could set off enormous fires. Similarly, if both took in oxygen and released carbon dioxide, life would eventually die out when all the oxygen had been used up.

In fact, the atmosphere is in a state of equilibrium in which, as Lovelock says, risk and benefit are nicely balanced.

Another finely-tuned aspect of our atmosphere is its density, which is ideally suited for us to breathe.

The Atmosphere and Respiration

We breathe every moment of our lives. We continuously take the air into our lungs and let it out. We do it so much that we might think of it as normal. In fact, respiration is quite a complex process.

Our bodily systems are so perfectly designed that we don't need to think about breathing. Our body estimates how much oxygen it needs and arranges for the delivery of the right amount whether we're walking, running, reading a book, or sleeping. The reason breathing is so important to us is that the millions of reactions that must constantly take place in our bodies to keep us alive all require oxygen.

Your ability to read this book is thanks to the millions of cells in the retina of your eye constantly being supplied with oxygen-derived energy. Similarly, all the tissues of our bodies and the cells forming them get their energy from the "burning" of carbon compounds in oxygen. The product of this burning–carbon dioxide–must be discharged from the body. If the level of oxygen in your bloodstream drops to low, the result is fainting; and if the absence of oxygen persists for more than a few minutes, the result is death.

And that's why we breathe. When we inhale, oxygen floods into about 300 million tiny chambers in our lungs. Capillary veins attached to these chambers absorb the oxygen in a twinkling and convey it first to heart and then to every other part of our body. The cells of our body use this oxygen and release carbon dioxide into the blood, which conveys it back to the lungs where it is expelled. The whole thing takes less than half a second: "clean" oxygen comes in and "dirty" carbon dioxide goes out.

You might be wondering why there are so many (300 million) of those little chambers in the lungs. They're there to maximize the surface area that is exposed to the air. They're carefully folded up to occupy as little space as possible; if they were unfolded, the result would be enough to cover a tennis court.

There is another point here that we need to keep in mind. The chambers of the lungs and the capillaries connecting to them are designed so small and perfectly in order to increase the rate at which oxygen and carbon dioxide are exchanged. But that perfect design depends on other factors: the density, viscosity, and pressure of air must all be right in order for the air to move properly in and out of our lungs.

At sea level, air pressure is 760 mm of mercury and its density is about 1 gram/liter. Again at sea level, its viscosity is nearly 50 times that of water. You might think these numbers unimportant but they are vital for our lives because, as Michael Denton notes:

> **The overall composition and general character of the atmosphere**–its density, viscosity, and pressure, etc–**must be very similar to what it is, particularly for air-breathing organisms.**[62]

When we breathe, our lungs use energy to overcome a force called "airway resistance". This force is the result of the resistance of air to movement. Owing to the physical properties of the atmosphere however, this resistance is weak enough that our lungs can take air in and let it out with a minimum expenditure of energy. If air resistance were higher, our lungs would be forced to work harder to enable us to breathe. This can be explained by an example. It easy to draw water into the needle of an injector but drawing honey in is much more difficult. The reason is that honey is denser than water and also more viscous.

If the density, viscosity, and pressure of air were higher, breathing would be as difficult as drawing honey into a needle. Someone might say "That's easy to fix. We'll just make the hole of the needle larger to increase the rate of flow." But if we did that in the case of the capillaries in the

lungs, the result would be to reduce the size of the area in contact with air, with the result that less oxygen and carbon dioxide would be exchanged in the same amount of time and the respiratory needs of the body would not be satisfied. In other words, the individual values of air's density, viscosity and pressure must all fall within certain limits in order for it to be breathable and those of the air we breathe do exactly that.

Michael Denton comments on this:

> **It is clear that if either the viscosity or the density of air were much greater,** the airway resistance would be prohibitive and **no conceivable redesign of the respiratory system would be capable of delivering sufficient**

> **oxygen to a metabolically active air-breathing organism...** By plotting all possible atmospheric pressures against all possible oxygen contents, it becomes clear that there is only one unique tiny area... where all the various conditions for life are satisfied... **It is surely of enormous significance that several essential conditions are satisfied in this one tiny region in the space of all possible atmospheres.**[63]

The numerical values of the atmosphere are not only necessary for us to breathe but are also essential for our Blue Planet to stay blue. If sea-level atmospheric pressure were much lower than its present value, the rate of water vaporization would be much higher. Increased water in the atmosphere would have a "greenhouse effect" trapping more heat and raising the average temperature of the planet. On the other hand, if the pressure were much higher, the rate of water vaporization would be less, turning large parts of the planet into desert.

All these finely-tuned equilibriums indicate that our atmosphere has been deliberately designed precisely so that life on Earth can exist. This is the reality discovered by science and it shows us again that the universe is not just an accidental jumble of matter. Undoubtedly there is a Creator ruling the universe, shaping matter as He wants it to be, and reigning over the galaxies, stars and planets under His sovereignty.

That supreme power, as the Qur'an tells us, is Allah, Lord of the whole universe.

And the Blue Planet on which we live is specially designed and **"smoothed out" by Allah for people** as stated in the Qur'an. (Surat an-Naziat 30) There are other verses revealing that Allah has created Earth for mankind to live in:

> **It is Allah who made the earth a stable home for you and the sky a dome, and formed you, giving you the best of forms, and provided you with good and wholesome things. That is Allah, your Lord. Blessed be Allah, the Lord of all the worlds. (Surah Ghafir: 64)**
>
> **It is He Who made the earth submissive to you, so walk its broad trails and eat what it provides. The Resurrection is to Him. (Surat al-Mulk: 15)**

The Equilibriums that Make Life Possible

The things we have mentioned so far are just a few of the delicate equilibriums that are essential for life on Earth. Examining the earth, we can make the list of the "essential factors for life" a long as we please. The American astronomer Hugh Ross has made a list of his own:

Surface Gravity;

- If stronger: atmosphere would retain too much ammonia and methane
- If weaker: planet's atmosphere would lose too much water

Distance From Parent Star;

- if farther: planet would be too cool for a stable water cycle
- if closer: planet would be too warm for a stable water cycle

Thickness of crust;

- if thicker: too much oxygen would be transferred from the atmosphere to the crust
- if thinner: volcanic and tectonic activity would be too great

Rotation period;

-If longer: diurnal temperature differences would be too great

-If shorter: atmospheric wind velocities would be too great

Gravitational interaction with moon;

- If greater: tidal effects on the oceans, atmosphere, and rotational period would be too severe
- If less: orbital obliquity changes would cause climatic instabilities

Magnetic Field;

- If stronger: electromagnetic storms would be too severe
- If weaker: inadequate protection from hard stellar radiation

Albedo (Ratio of Reflected light to total amount falling on surface);

- If greater: runaway ice age would develop
- If less: runaway greenhouse effect would develop

Oxygen to nitrogen ratio in the atmosphere;

- if larger: advanced life functions would proceed too quickly
- if smaller: advanced life functions would proceed too slowly

Carbon dioxide and water vapour levels in atmosphere;

- if greater: runaway greenhouse effect would develop
- if less: greenhouse effect would be insufficient

Ozone level in Atmosphere;

- if greater: surface temperature would be too low
- if less: surface temperatures would be too high; there would be too much uv radiation at the surface

Seismic Activity;

- if greater: too many life-forms would be destroyed
- if less: nutrients on ocean floors (from river runoff) would not be recycled to the continents through tectonic uplift.[64]

These are just some of the "design decisions" that had to be made in order for life to exist and survive. But even these are enough to show that the earth did not come into being as a result of chance nor was it formed as a result of a lucky chain of events.

These and a myriad other details reaffirm a plain and simple truth: Allah and Allah alone created the universe, the stars, planets, mountains, and seas perfectly, giving life to human beings and other living things, and placing His creations under the control of mankind. Allah and Allah alone, the source of mercy and might, is powerful enough to create something from nothingness.

This perfect creation of Allah is described in the Qur'an thus:

Are you stronger in structure or is heaven? He built it. He raised its vault high and made it level. He darkened its night and brought forth its morning light. After that He smoothed out the earth and brought forth from it its water and its pastureland and made the mountains firm for you and for your livestock to enjoy. (Surat an-Nazi'at: 27-33)

CHAPTER VI

THE DESIGN IN LIGHT

That the radiation from the sun (and from many sequence stars) should be concentrated into a minuscule band of the electromagnetic spectrum which provides precisely the radiation required to maintain life on earth is very remarkable.

Ian Campbell, British Physicist[65]

The sun is probably the one thing we see most often throughout our lives. Whenever we raise our sight to the sky during the day, we can see its dazzling light. If someone were to come up and ask "What good is the sun? we would probably reply without even a thought that the sun gives us light and heat. That answer, although a bit superficial, would be correct.

But does the sun just "happen" to radiate light and heat for us? Is it accidental and unplanned? Or is the sun specially designed for us? Could this great ball of fire in the sky be a gigantic "lamp" that was created so as to meet our exact needs?

Recent research indicates that the answer to the last two questions is "yes". "Yes" because in sunlight there is a design that inspires amazement.

The Right Wavelength

Both light and heat are different manifestations of electromagnetic radiation. In all its manifestations, electromagnetic radiation moves through space in waves similar to those created when a stone is thrown into a lake. And just as the ripples created by the stone may have different heights and the distances between them may vary, electromagnetic radiation also has different wavelengths.

The analogy shouldn't be taken too far however because there are huge differences in the wavelengths of electromagnetic radiation. Some are several kilometers long while others are shorter than a billionth of a centimeter and the other wavelengths are to be found in a smooth, unbroken spectrum everywhere in between. To make things easier, scientists divide this spectrum up according to wavelength and they assign different names to different parts of it. The radiation with the shortest wavelength (one-trillionth of a centimeter) for example is called "gamma rays": these rays pack tremendous amounts of energy. The longest wavelengths are called "radio waves": they can be several kilometers long but carry very little energy. (One result of this is that radio waves are quite harmless to us while exposure to gamma rays can be fatal.) Light is a form of

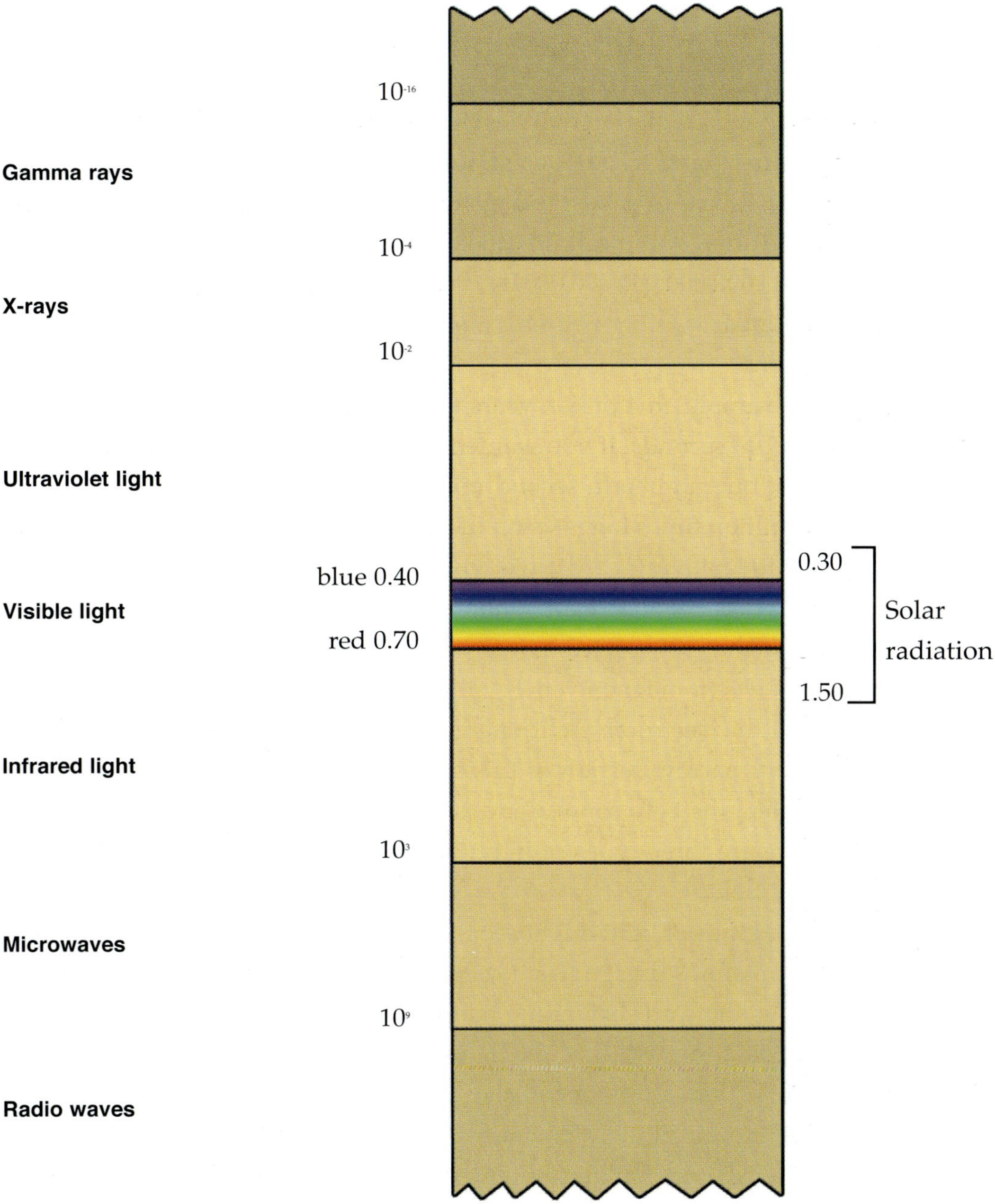

THE DIFFERENT WAVELENGTHS OF ELECTROMAGNETIC RADIATION

The stars and other sources of light in the universe do not all give out the same kind of radiation. Instead, they radiate energy with a broad range of wavelengths. Gamma rays, which have the shortest wavelengths, are just $1/10^{25}$ the length of the longest radio waves. Strangely enough, nearly all of the radiation emitted by the sun falls into a single band that is also $1/10^{25}$ of the whole spectrum. The reason, is that the only kinds of radiation that are necessary and fit for life fall in this narrow band.

electromagnetic radiation that lies between these two extremes.

The first thing to be noticed about the electromagnetic spectrum is how broad it is: the longest wavelength is 10^{25} times the size of the shortest one. Written out in full, 10^{25} looks like this:

10,000,000,000,000,000,000,000,000

A number that big is pretty meaningless by itself. Let's make a few comparisons.

For example, in 4 billion years (the estimated age of the earth) there are about 10^{17} seconds. If you wanted to count from 1 to 10^{25} and did so at the rate of one number a second nonstop, day and night, it would take you 100 million times longer than the age of the earth! If we were to build a pile of 10^{25} playing cards, we would end up with a stack stretching halfway across the observable universe.

This is the vast spectrum over which the different wavelengths of the universe's electromagnetic energy extend. Now the curious thing about this is that the electromagnetic energy radiated by our sun is restricted to a very, very narrow section of this spectrum. 70% of the sun's radiation has wavelengths between 0.3 and 1.50 microns and within that narrow band there are three types of light: visible light, near-infrared light, and ultraviolet light.

Three kinds of light might seem quite enough but all three combined make up an almost insignificant section of the total spectrum. Remember our 10^{25} playing cards extending halfway across the universe? Compared with the total, the width of the band of light radiated by the sun corresponds to just one of those cards!

Why should sunlight be limited to such a narrow range?

The answer to that question is crucial because the only radiation that is capable of supporting life on earth is the kind that has wavelengths falling within this narrow range.

In *Energy and the Atmosphere*, the British physicist Ian Campbell addresses this question and says "**That the radiation from the sun (and from many sequence stars) should be concentrated into a minuscule band of the electromagnetic spectrum which provides precisely the radiation required to**

maintain life on earth is very remarkable." According to Campbell, this situation is "staggering".[66]

Let us now examine this "staggering design of light" more closely.

From Ultraviolet to Infrared

We said that there was a range of $1:10^{25}$ in the sizes of the longest and shortest electromagnetic wavelengths. We also said that the amount of energy that was carried depended upon the wavelength: shorter wavelengths pack more energy than longer ones. Another difference has to do with how radiation at different wavelengths interacts with matter.

The shortest forms of radiation are called (in increasing order of wavelength) "gamma rays", "X-rays", and "ultraviolet light". They have the ability to split atoms because they are so highly energized. All three can cause molecules–especially organic molecules–to break up. In effect, they tear matter apart at the atomic or molecular level.

Radiation with wavelengths longer than visible light begins at infrared and extends up to radio waves. Its impact upon matter is less serious because the energy it conveys is not as great.

The "impact upon matter" that we spoke of has to do with chemical reactions. A significant number of chemical reactions can take place only if energy is added to the reaction. The energy required to start a chemical reaction is called its "energy threshold". If the energy is less than this threshold, the reaction will never start and if it is more, it is of no good: in either case, the energy will have been wasted.

In the whole electromagnetic spectrum, there is just one little band that has the energy to cross this threshold exactly. Its wavelengths range between 0.70 microns and 0.40 microns and if you'd like to see it, you can: just raise your head and look around–it's called "visible light". This radiation causes chemical reactions to take place in your eyes and that is why you are able to see.

The radiation known as "visible light" makes up 41% of sunlight even though it occupies less than $1/10^{25}$ of the whole electromagnetic

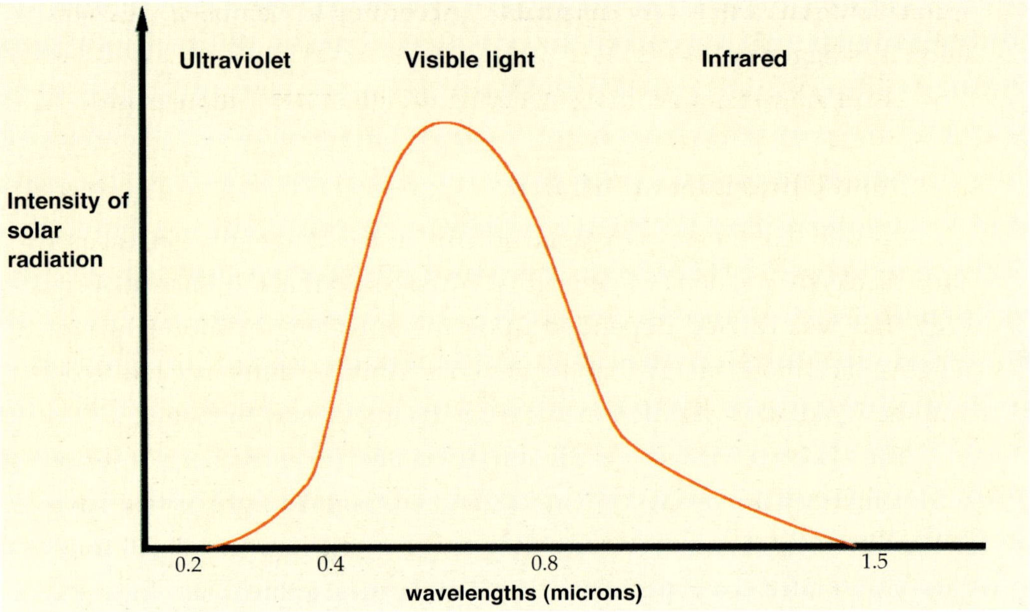

Nearly all of the sun's radiation is restricted to a narrow band of wavelengths ranging from 0.3 to 1.50 microns. This band encompasses near ultraviolet, visible, and infrared light.

spectrum. In his famous article "Life and Light", which appeared in *Scientific American,* the renowned physicist George Wald considered this matter and wrote **"the radiation that is useful in prompting orderly chemical reactions comprises the great bulk of that of our sun."**[67] That the sun should radiate light so exactly right for life is indeed an extraordinary example of design.

Is the rest of the light the sun radiates good for anything?

When we look at this part of the light we see that a large part of solar radiation falling outside the range of visible light is in the section of the spectrum called **"near infrared"**. This begins where visible light ends and again occupies a very small part of the total spectrum–less than $1/10^{25}$.[68]

Is infrared light good for anything? Yes, but this time it's no use to look around because you can't see it with the naked eye. However you can easily feel it: the warmth you feel on your face when you look up on a bright sunny summer or spring day is caused by infrared radiation coming from the sun.

The sun's infrared radiation is what carries the thermal energy that keeps Earth warm. It too is as essential for life as visible light is. And the fascinating thing is that our sun was apparently created just to serve for these two purposes, because these two kinds of light make up the greatest part of sunlight.

And the third part of sunlight? Is that of any benefit?

You can bet on it. This is **"near ultraviolet light"** and it makes up the smallest fraction of sunlight. Like all ultraviolet light, it is highly energized and it can cause damage to living cells. The sun's ultraviolet light however is the "least harmful" kind since it is closest to visible light. Although overexposure to solar ultraviolet light has been shown to cause cancer and cellular mutations, it has one vital benefit: the ultraviolet light concentrated in such a miniscule band[69] is needed for the synthesis of vitamin D in humans and other vertebrates. (Vitamin D is necessary for the formation and nourishment of bone: without it, bones become soft or malformed, a disease called rickets that occurs in people deprived of sunlight for great lengths of time.)

In other words, all the radiation emitted by the sun is essential to life: none of it is wasted. The amazing thing is that all this radiation is limited to a $1/10^{25}$ interval of the whole electromagnetic spectrum yet it is sufficient to keep us warm, see, and allow all the chemical reactions necessary for life to take place.

Even if all the other conditions necessary for life and mentioned elsewhere in this book existed, if the light radiated by the sun fell into any other part of the electromagnetic spectrum, there could be no life on Earth. It is certainly impossible to explain the fulfillment of this condition having a probability of 1 in 10^{25} with a logic of coincidence.

And if all this were not enough, light does something else: it keeps us fed, too!

Photosynthesis and Light

Photosynthesis is a chemical process whose name almost everyone who's ever gone to school will be familiar with. Most people however fail to realize how vitally important this process is for life on Earth or what a mystery its workings are.

First let's brush off our high-school chemistry and take a look at the formula for the photosynthesis reaction:

$$6H_2O + 6CO_2 + \text{Sunlight} \dashrightarrow \underset{\text{Glucose}}{C_6H_{12}O_6} + 6O_2$$

Translated into words this means: Water and carbon dioxide and sunlight produces glucose and oxygen.

To be more exact what is happening in this chemical reaction is that six molecules of water (H_2O) combine with six molecules of carbon dioxide (CO_2) in a reaction that is energized by sunlight. When the reaction is complete, the result is a single molecule of glucose ($C_6H_{12}O_6$), a simple sugar that is a fundamental element of nutrition-, and six molecules of gaseous oxygen (O_2). The source of all nutriments on our planet, glucose contains a great deal of energy.

Simple though this reaction may look, it is in fact incredibly complex. There is only one place where it occurs: in plants. The plants of this world produce the basic food for all living things. Every other living thing is ultimately nourished in one way or another by glucose. Herbivorous animals eat the plants themselves and carnivorous animals eat plants and/or other animals. Human beings are no exception: our energy is derived from the food we eat and comes from the same source. Every apple, potato, chocolate, or steak or anything else you eat is supplying you with energy that came from the sun.

But photosynthesis is important for another reason. The reaction has two products: in addition to glucose, it also releases six molecules of oxygen. What's happening here is that plants are continuously cleaning up an

For hundreds of millions of years, plants have been busy doing something no laboratory has ever been able to duplicate: Using sunlight, the produce food. A crucial condition for this extraordinary transformation however is that the light that the plants receive must be precisely right for photosynthesis to take place.

atmosphere that is constantly being "polluted" by air-breathing creatures-human beings and animals, whose energy is derived from combustion in oxygen, a reaction that produces carbon dioxide. If plants didn't release oxygen, the oxygen-breathers would eventually use up all the free oxygen in the atmosphere and that would be the end of them. Instead, the oxygen in the atmosphere is constantly being replenished by plants.

Without photosynthesis, plant life could not exist; and without plant life, there would be no animal or human life. This marvelous chemical reaction, which has never been duplicated in any laboratory, is taking place deep in the grass you step on and in trees you may not even notice. It once occurred in the vegetables on your dinner plate. It is one of the fundamental processes of life.

The interesting thing is what a carefully-designed process photosynthesis is. When we study it, we can't help but observe that there is a perfect balance between plant photosynthesis and the energy consumption of oxygen-breathers. Plants supply glucose and oxygen. Oxygen-breathers burn the glucose in the oxygen in their cells to get energy and they release carbon dioxide and water (in effect, they're reversing the photosynthesis reaction) that the plants use to make more glucose and oxygen. And so it goes on, a continuous cycle that is called the "carbon cycle" and it is powered by the energy of the sun.

In order to see how perfectly-created this cycle truly is, let us focus our attention on just one of its elements for the moment: the sunlight.

In the first part of this chapter we looked at sunlight and found that its radiation components were specially tailored to allow life on Earth. Could sunlight also be deliberately tailored for photosynthesis as well? Or are plants flexible enough so that they can perform the reaction no matter which kind of light reaches them?

The American astronomer George Greenstein discusses this in *The Symbiotic Universe*:

> Chlorophyll is the molecule that accomplishes photosynthesis... The mechanism of photosynthesis is initiated by the absorption of sunlight by a chlorophyll molecule. But in order for this to occur, the light must be of the right color. Light of the wrong color won't do the trick.
>
> A good analogy is that of a television set. In order for the set to receive a given channel it must be tuned to that channel; tune it differently and the reception will not occur. It is the same with photosynthesis, the Sun functioning as the transmitter in the analogy and the chlorophyll molecule as the receiving TV set. If the molecule and the Sun are not tuned to each other-tuned in the sense of colour- photosynthesis will not occur. **As it turns out, the sun's color is just right**.[70]

In the last chapter we drew attention to the error inherent in the idea of the adaptability of life. Some evolutionists hold that "if conditions had been different, life would have evolved to be perfectly in harmony with them as well". Thinking superficially about photosynthesis and plants, one could come to a similar conclusion: "If sunlight were different, plants would have just evolved according to that." But this is in fact impossible.

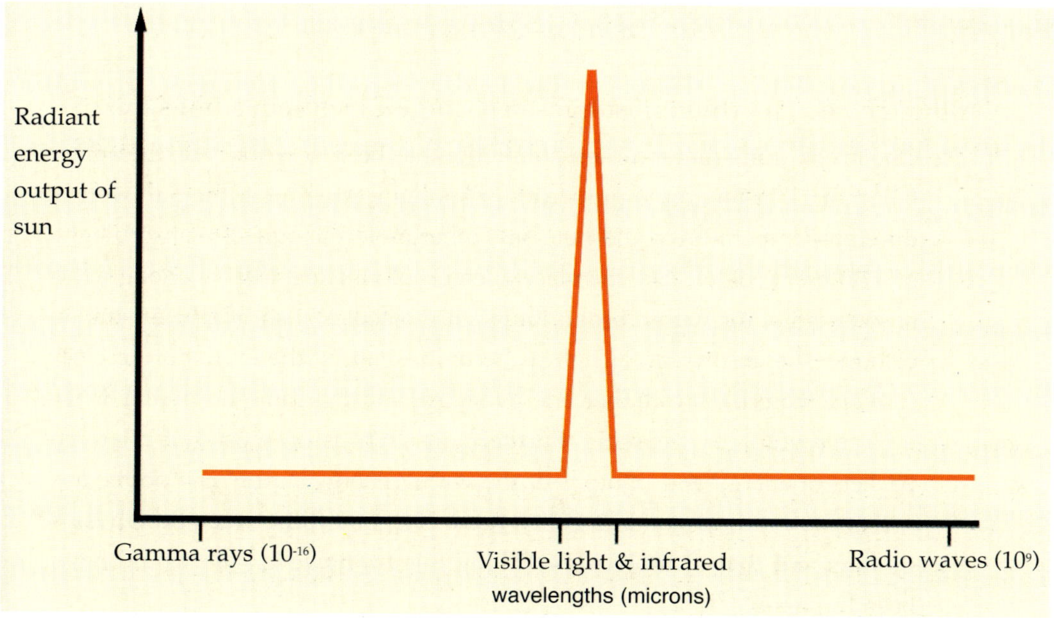

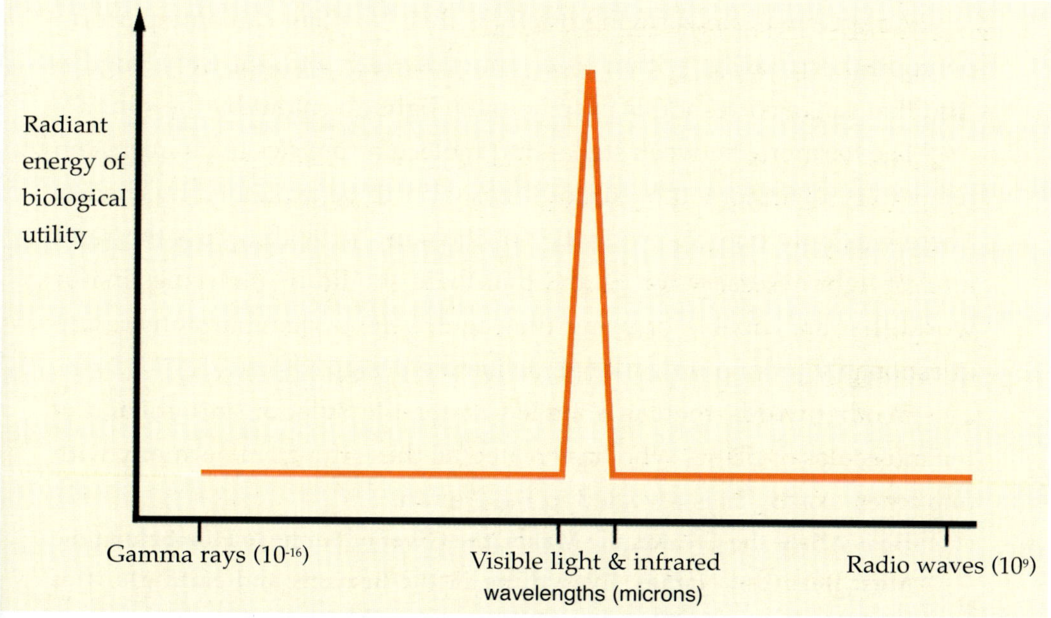

THE FITNESS OF SUNLIGHT AND CHLOROPHYLL

Plants are able to perform photosynthesis because the chlorophyll molecules in their cells are sensitive to sunlight. But chlorophyll is only able to use a very limited range of light wavelengths and those are the wavelengths that the sun radiates the most. What is even more interesting is that this interval corresponds to just $1/10^{25}$ of the whole electromagnetic spectrum.

In the two graphs above, the extraordinary fitness between sunlight and chlorophyll can be seen. In the upper chart is the distribution of the light emitted by the sun. In the lower one is the light under which photosynthesis will work. The fact that these two curves are almost identical is an indication of how perfectly designed visible light is.

Although he's an evolutionist himself, George Greenstein admits this:

> One might think that a certain adaptation has been at work here: the adaptation of plant life to the properties of sunlight. After all, if the Sun were a different temperature could not some other molecule, tuned to absorb light of a different colour, take the place of chlorophyll? Remarkably enough the answer is no, for within broad limits all molecules absorb light of similar colours. The absorption of light is accomplished by the excitation of electrons in molecules to higher energy states, and the same no matter what molecule you are discussing. Furthermore, light is composed of photons, packets of energy and photons of the wrong energy simply can not be absorbed... As things stand in reality, **there is a good fit between the physics of stars and that of molecules.** Failing this fit, however, life would have been impossible.[71]

What Greenstein is saying briefly is this: No plant can only perform photosynthesis except within a very narrow range of light wavelengths. And that range corresponds exactly to the light given out by the sun.

The harmony between stellar and molecular physics that Greenstein refers to is a harmony too extraordinary ever to be explained by chance. There was only one chance in 10^{25} of the sun's providing just the right kind of light necessary for us and that there should be molecules in our world that are capable of using that light. This perfect harmony is unquestionably proof of intentional, deliberate design.

In other words, there is a single Creator, the Ruler of starlight and of the molecules of plants Who has created all these things in harmony with one other, exactly as is revealed in the Qur'an:

> **He is Allah- the Creator, the Maker, the Giver of Form. To Him belong the Most Beautiful Names. Everything in the heavens and earth glorifies Him. He is the Almighty, the All Wise. (Surat al-Hashr: 24)**

The Light of Your Eyes

We have seen how the light coming to us from the sun consists of just three narrow bands of the electromagnetic spectrum:

1) Infrared light, whose wavelengths are longer than visible light and

which keeps Earth warm.

2) A small amount of ultraviolet light, whose wavelengths are shorter than visible light and which is necessary for the synthesis of vitamin D among other things.

3) Visible light, which makes vision possible and supports plant photosynthesis.

The existence of a range of "visible light" is as important for the support of biological vision as it is for photosynthesis. The reason is that it is impossible for a biological eye to see any band of the spectrum outside that of visible light and a very small section of near infrared.

To explain why this should be so, we first need to understand how vision takes place. It begins with particles of light called "photons" passing through the pupil of eye and falling onto the surface of the retina located at the back of the eye. The retina contains cells that are light-sensitive. They are so sensitive that each can recognize when even a single photon strikes it. The photon's energy activates a complex molecule called "rhodopsine", large quantities of which are contained in these cells. The rhodopsine in turn activates other cells and those activate still others in turn.[72] Eventually an electrical current is generated and this is carried to the brain by the optic nerves.

The first requirement for this system to work is that the retina cell must be able to recognize when a photon strikes it. For that to happen, the photon must carry an exact amount of energy: if it is too much or too less, it won't activate the formation of rodopsine. Changing the size of the eye makes no difference: the crucial thing is the harmony between the size of the cell and the wavelengths of the photons coming in.

Designing an organic eye that could see other ranges of the electromagnetic spectrum turns out to be impossible in a world dominated by carbon-based life. In *Nature's Destiny*, Michael Denton explains this subject in detail and confirms that an organic eye can only see within the range of visible light. While other models of eyes that could, in theory, be designed, none of them would be able to see different ranges of the spectrum. Denton tells us why:

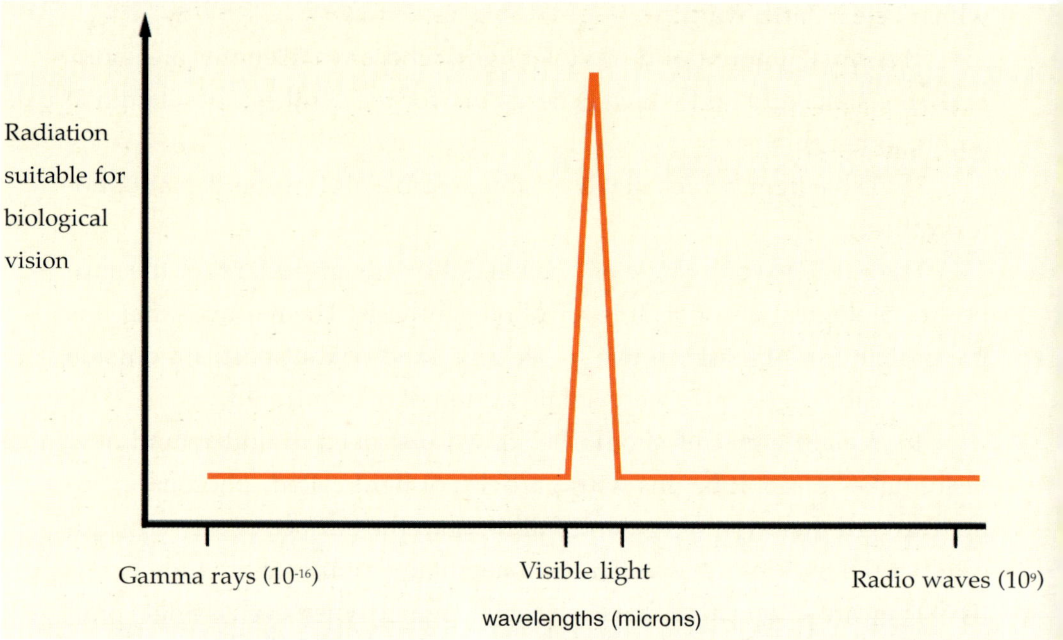

The only rays of light that are suitable for biological vision have wavelengths that fall within the range of what is called "visible light". A large part of the energy that is emitted by the sun falls in that range.

> UV, X-ray, and gamma rays are too energetic and are highly destructive, while infrared and radio waves are too weak to be detected because they impart so little energy interacting with matter... And so it would appear that for several different reasons, the visual region of the electromagnetic spectrum is the one region supremely fit for biological vision and particularly for the high-resolution vertebrate camera eye of a design and dimension very close to that of the human eye.[73]

Pausing to think about everything that has been said so far, we come to this conclusion: The sun radiates energy within a narrow band (a band so narrow that it corresponds to just $1/10^{25}$ of the whole electromagnetic spectrum) that has been carefully chosen. So finely adjusted is this band that it keeps the world warm, supports the biological functions of complex life-forms, enables photosynthesis, and allows the creatures of this world to see.

The Right Star, the Right Planet, and the Right Distance

In "The Blue Planet" we compared our world with the other planets of the solar system and found that the range of temperatures necessary for life exists only on Earth. The biggest reason for this is that the earth is just the right distance from the sun: the outer planets like Mars, Jupiter, or Pluto are too cold while the inner planets Venus and Mercury are too hot.

Those who refuse to admit that there is intentional design in the distance between Earth and Sun suggest something like the following:

"The universe is full of stars, some of them much bigger than the sun and some of them much smaller. These could very well have planetary systems of their own. If a star is bigger than the sun, then the ideal planet for life would be located at a much greater distance than the earth is from the sun. For example, a planet in an orbit around a red giant at the distance of Pluto could have a temperate climate like our world has. Such a planet would be just as fit for life as our earth is."

The claim is invalid in one very important respect for it ignores the fact that stars of different masses radiate different types of energy.

The factors that determine the wavelengths of the energy that a star radiates are its mass and its surface temperature (the latter of which is directly related to mass). For example, the sun radiates near ultraviolet, visible, and near infrared light because its surface temperature is around 6,000°C. If the sun's mass were a bit bigger, its surface temperature would be higher; but in that case, the energy levels of the sun's radiation would also be higher and the sun would be radiating much more destructive ultraviolet rays than it does.

This tells us that any star that is to radiate light that will support life absolutely must have a mass close to that of our sun. But if there are to be life-supporting planets orbiting around such stars, those planets must be located at distances not substantially different from that between the earth and the sun.

In other words, no planet revolving around a red giant, a blue giant, or any other star whose mass was substantially different from the sun's

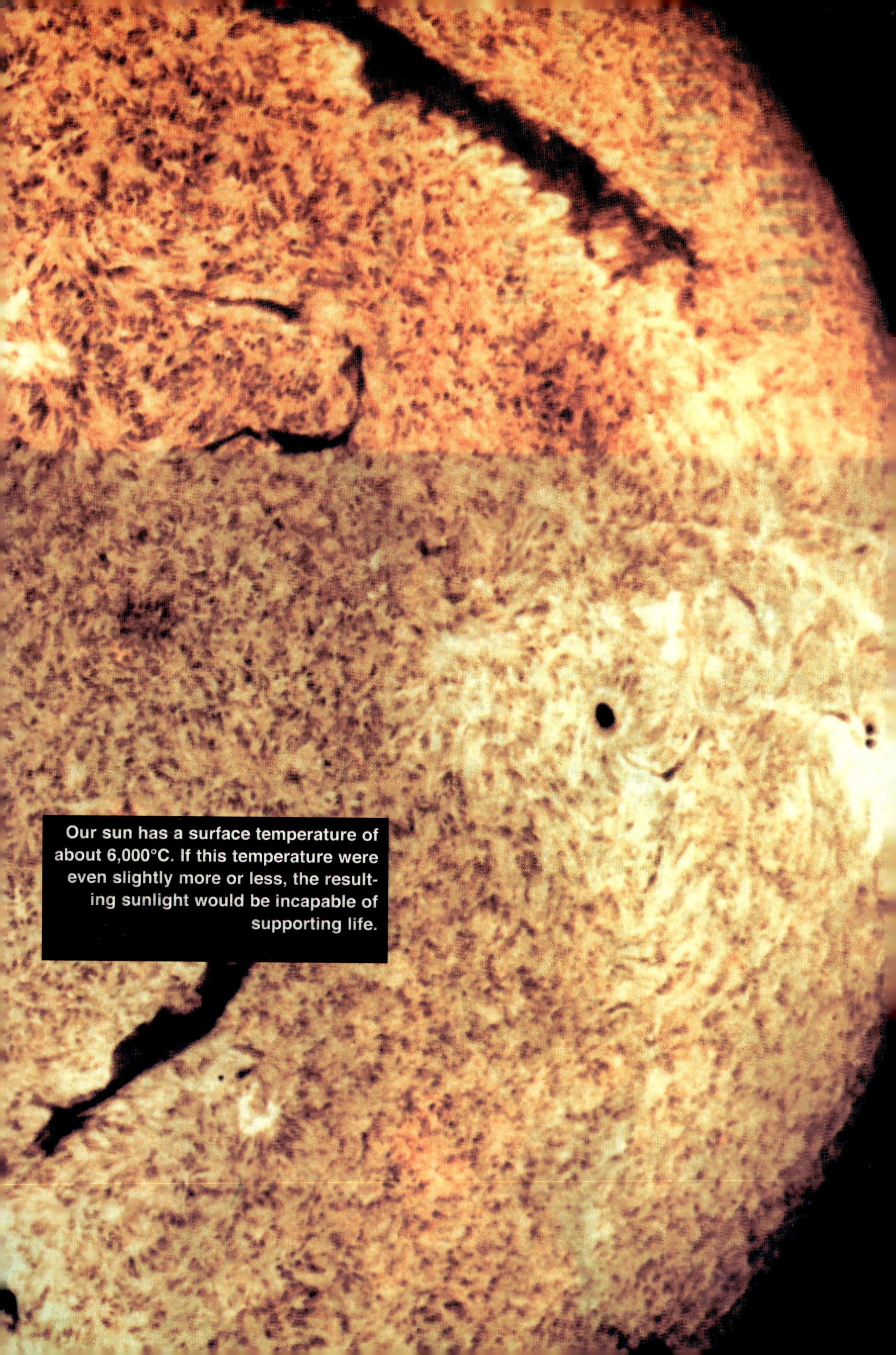

Our sun has a surface temperature of about 6,000°C. If this temperature were even slightly more or less, the resulting sunlight would be incapable of supporting life.

could harbor life. The only source of energy capable of supporting life is a star like our sun. **The only planetary distance that is suitable for life is the distance between the earth and the sun.**

There is another way of expressing this truth: The sun and the earth were each created to be just as they needed to be. And indeed, in the Qur'an it is revealed that Allah created everything according to precise calculation:

> **It is He Who splits the sky at dawn, and appoints the night as a time of stillness and the sun and moon as a means of reckoning. That is what the Almighty, the All-Knowing has ordained. (Surat al-Anam: 96)**

The Harmony of Light and Atmosphere

Since the beginning of this chapter we have been talking about the radiation given out by the sun and how it was specially designed to support life. There is yet another crucially important factor that we have not yet touched upon: In order for this radiation to reach the earth's surface, it has to pass through the atmosphere.

Sunlight certainly couldn't do us any good if the atmosphere didn't let it through. But it does; in fact, our atmosphere is specially designed to be transparent to this beneficial radiation.

The really interesting thing is not so much that the atmosphere allows beneficial sunlight to pass but that sunlight is the only radiation that it allows through. The atmosphere lets in the visible and near infrared light that is necessary for life but it blocks other forms of radiation that are deadly. This makes the atmosphere an important filter against the cosmic radiation that reaches the earth from the sun and from other sources. Denton has this to say about the matter:

> Atmospheric gases themselves absorb electromagnetic radiation immediately on either side of the visible and near infrared... The only region of the spectrum allowed to pass through the atmosphere over the entire range of electromagnetic radiation from radio to gamma rays is the exceedingly narrow band including the visible and near infrared. Virtually no gamma, X, ultraviolet, far infrared, and microwave radiation reaches the surface of the earth.[74]

It is impossible to ignore the artfulness of this design. The sun sends only $1/10^{25}$ of the whole range of electromagnetic radiation that could be sent, that happens to be the range that is good only for us, and that is the radiation that the atmosphere lets through! At this point it's also worth pointing out that nearly all of the near ultraviolet that the sun radiates gets trapped by the atmosphere's ozone layer.

Another point that makes this even more interesting is that, like air, water also has an extremely particular sort of transparency: the only radiation capable of spreading through water is the range of visible light. Even near infrared radiation, which penetrates the atmosphere (and thus provides heat) penetrates only a few millimeters into water. Because of this, only a few millimeters of the surface of the world's oceans are heated by radiation from the sun. That heat is conveyed in stages to lower levels and as a result of this, below a particular depth, the temperature of the seawater is quite similar all over the world. This of course creates an environment quite suitable for life.

Another interesting point concerning water is that the different colors of visible light are able to travel different distances in it. Below eighteen meters, for example, red light cannot penetrate while yellow can reach depths of up to a hundred meters. Blue and green on the other hand descend to 240 meters. This is an extremely important design because the light that is particularly crucial for photosynthesis is the blue and green portion of the spectrum. Since water allows these colors to penetrate more deeply than the others, photosynthesizing plants can live up to 240 meters beneath the surface.

These are all facts of the utmost importance. No matter what physical law related to light we examine, we discover that everything has been exactly arranged so that life can exist. Commenting on this situation, *Encyclopedia Britannica* admits how extraordinary it all is:

> Considering the importance of visible sunlight for all aspects of terrestrial life, one can not help being awed by the dramatically narrow window in the atmosphere absorption and in the absorption spectrum of water. [75]

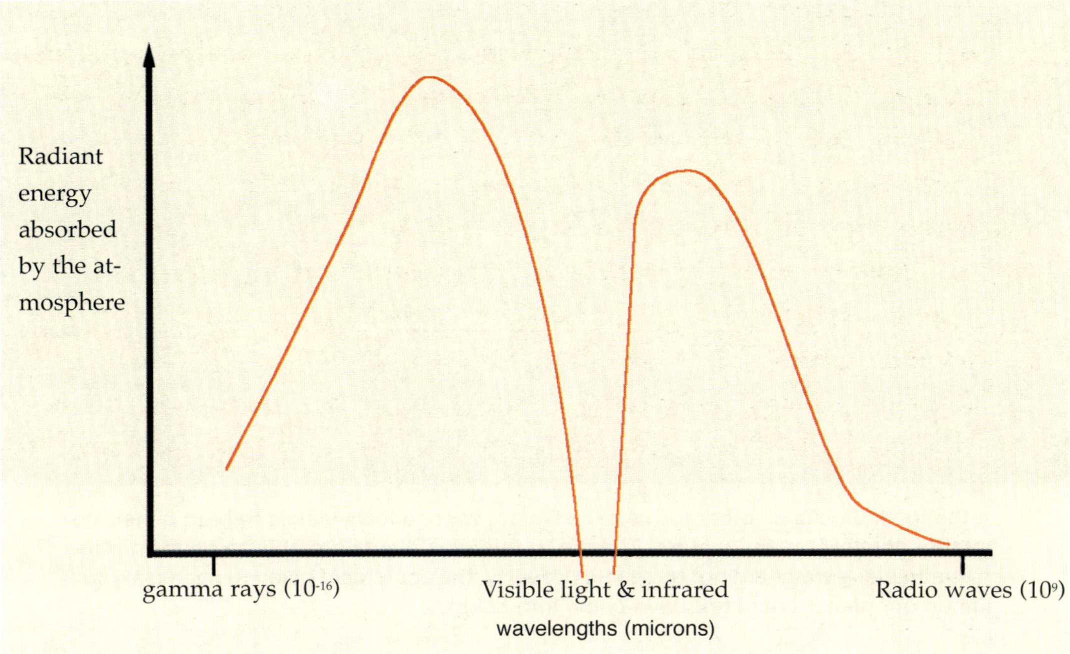

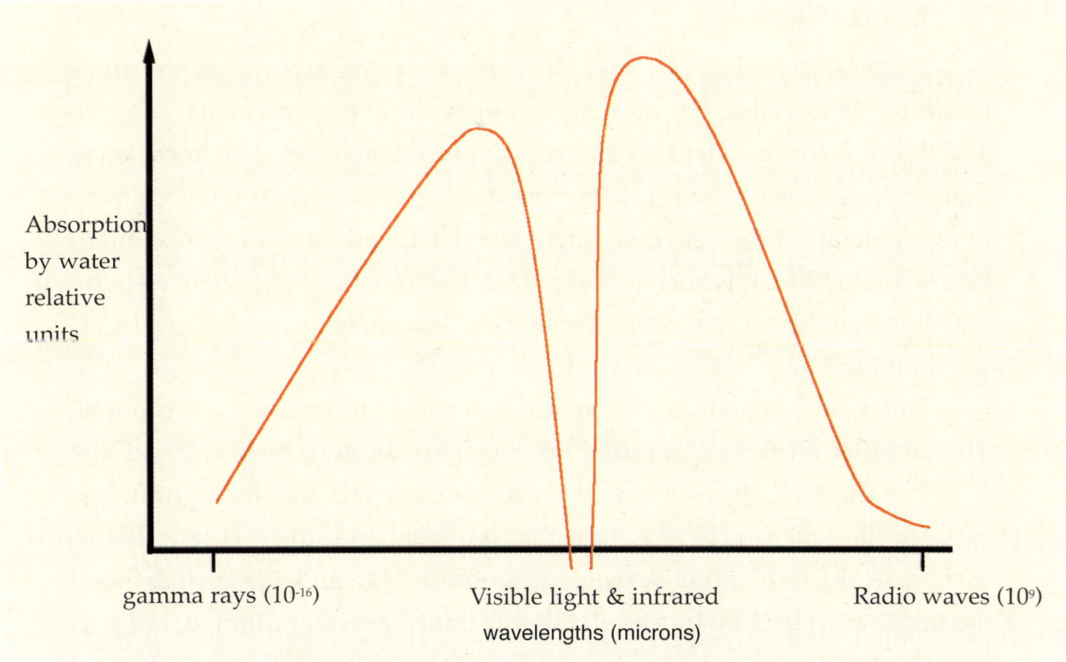

Air as well as water allows the passage of only that radiation that is necessary for us to live. All the harmful and deadly cosmic radiation coming from distant space is caught in this perfectly-designed filter.

Although it blocks all other forms of radiation, water allows visible light to penetrate into its depth for many meters. Thanks to this, sea plants are able to perform photosynthesis. If water did not have this property, the ecological balance necessary for life on our planet could not have come into being.

Conclusion

Materialist philosophy and Darwinism, which takes materialism as its source, both claim that human life appeared in the universe by chance and that it is an "accident" with no purpose whatsoever. The knowledge that is being gained through advances in science however is showing that, in every detail of the universe, there is a design and a plan whose intention is human life. It is such a design that, even such a component as light, which we might never have thought about before, is so clearly "just right" that one can't help but be amazed.

To try and explain such careful design as "accidental" is irrational. The fact that all the sun's radiation is constricted to a narrow band just $1/10^{25}$ of the total electromagnetic spectrum, the fact that the light necessary for life falls precisely within that narrow band, the fact that the atmosphere blocks all other wavelengths of radiation and admits just these, the fact that water also blocks all other forms of deadly radiation and permits the passage only of visible light: Can these really all be coincidences? Such extraordinary fine-tuning as this can be explained not by chance but only by conscious design. This in turn shows us that the whole universe and all the details of that universe–including the light of the sun that enables us to see and keeps us warm–have been specially created and arranged for us to live.

The conclusion reached by science is a truth that has been taught to mankind in the Qur'an for fourteen centuries. Science shows that sunlight has been created for us, in other words, that it has been made to be "at our service". In the Qur'an we are told that **"The sun and moon both run with precision."** (Surat ar-Rahman: 5) Elsewhere it is stated:

> **Allah is He who created the heavens and the earth and sends down water from the sky and by it brings forth fruits as provision for you. ...He has made the sun and moon subservient to you holding steady to their courses, and He has made the night and day subservient to you. He has given you everything you have asked Him for. If you tried to number Allah's blessings, you could never count them. Man is indeed wrongdoing, ungrateful. (Surah Ibrahim: 32-34)**

CHAPTER VII

THE DESIGN IN WATER

This, as most other of the Atheists' Arguments, proceeds from a deep Ignorance of Natural Philosophy; for if there were but half the sea that now is, there would also be but half the Quantity of Vapours, and consequently we could have but half as many Rivers as now there are to supply all the dry land we have at present, and half as much more; for the quantity of Vapours which are raised, as well as to the heat which raised them. The Wise Creator therefore did so prudently order it, that the seas should be large enough to supply Vapours sufficient for all the land.

John Ray, 18th century British Naturalist[76]

Most of our planet is covered with water. Oceans and seas make up three fourths of the earth's surface while the land itself contains countless numbers of rivers and lakes. The snow and ice on the summits of lofty mountains is water in its frozen form. A substantial part of the earth's water is in the sky: every cloud contains thousands--sometimes millions–of tons of water in the form of vapor. From time to time some of this water vapor turns into drops of liquid and falls to the ground: in other words, it rains. Even the air you're breathing now contains a certain amount of water vapor.

In short, no matter where you may look on the surface of the earth, you're certain to see water around somewhere. Indeed, the room you're sitting in at this moment probably contains about forty to fifty liters of water in it. Look around. You can't see it? Look again, more carefully, this time raising your eyes from these words and look at your hands, arms, legs, and body. That 40-50 liter mass of water is you!

It's you because about 70% of the human body is water. Your body's cells contain many things but nothing so much or so important as water. The biggest part of the blood that circulates everywhere in your body is of course water. This is true not just of yourself or of other people however: the bulk of the bodies of all living things is water. Without water it seems, life is impossible.

Water is a substance that was specially designed so as to be the basis of life. Each and every one of its physical and chemical properties was specially created for life.

The Fitness of Water

The biochemist A. E. Needham notes how essential liquids are for life to form in his book *The Uniqueness of Biological Materials*. If the laws of the universe had allowed only solids or gases to exist, there never would have been any life. The reason is that the atoms of solids are too tightly-packed and static and simply will not allow the dynamic molecular processes that are necessary for life to take place. In gases, on the other

hand, the atoms move about freely and chaotically: it would be impossible for the complex mechanisms of life-forms to function within such a structure.

In short, the existence of a liquid environment is essential in order for the processes necessary for life to take place. The most ideal of all liquids–or rather, the only ideal liquid–for this purpose is water.

That water possesses properties that are extraordinarily fit for life is something that drew the attention of scientists long ago. The first attempt to investigate this subject in detail however was *Astronomy and General Physics Considered with Reference to Natural Theology,* a book by the English naturalist William Whewell that was published in 1832. Whewell had been examining the thermal properties of water and noticed that some of them seemed to violate the accepted rules of natural law. The conclusion he drew from this was that these inconsistencies should be taken as proof that this substance had been specially created in order for life to exist.

The most comprehensive analysis of the suitability of water for life was to come from Lawrence Henderson, a professor in the Department of

Biological Chemistry of Harvard University, about a century after Whewell's book. In his book *The Fitness of the Environment,* which some were later to call "the most important scientific work of the first quarter of the 20th century", Henderson reaches this conclusion concerning the natural environment of our world:

> The fitness...(of these compounds constitutes) a series of maxima-unique or nearly unique properties of water, carbon dioxide, the compounds of carbon, hydrogen, and oxygen and the ocean - so numerous, so varied, so complete among all things which are concerned in the problem that together they form certainly the greatest possible fitness.[77]

The Extraordinary Thermal Properties of Water

One of the subjects dealt with in Henderson's book is the thermal properties of water. Henderson notes that there are five distinct ways in which the thermal properties of water are unusual:

1) All known solids decrease in size as they grow colder. This is true of all known liquids as well: as their temperatures decrease, they lose volume. As volume decreases, density increases and thus the colder parts of the liquid become heavier. This is why the solid forms of substances weigh more (by volume) than they when they are in liquid form. There is one case where this "law" is violated: water. Like other liquids, water contracts in volume as it grows colder but it only does this down to a certain temperature (4°C) thereafter–unlike all other known liquids–it suddenly begins to expand and when it finally solidifies (freezes) it expands even more. As a result, "solid water" is lighter than "liquid water". According to the normal laws of physics, solid water, which is to say ice, ought to be heavier than liquid water and should sink to the bottom when it forms; instead, it floats.

2) When ice melts or water vaporizes, it absorbs heat from its surroundings. When these transitions are reversed (that is, when water freezes or vapor precipitates) heat is released. In physics the term "latent heat" is used to describe this.[78] All liquids have a latent heat of some sort

or other but that of water is among the highest known. At "normal" temperatures, the only liquid whose latent heat when freezing is superior to that of water is ammonia. In terms of its latent heat properties at vaporization on the other hand, no other liquid can compare with water.

3) The "thermal capacity" of water, that is, the amount of heat necessary to raise the temperature of water by one degree, is higher than the great majority of other liquids.

4) The thermal conductivity of water, its ability to convey heat, is at least four times higher than any other liquid.

5) The thermal conductivity of ice and snow on the other hand is low.

By now you are probably wondering what importance these seemingly technical five physical properties could possibly have. As it turns out, the significance of each and every one of them is enormous because life in general and our own life in particular is possible in this world just because these five properties are what they are.

Let's now take a look at them one by one.

The Effect of "Top-down" Freezing

Other liquids freeze from the bottom up; water freezes from the top down. This is the first unusual property of water mentioned above and it is crucial for the existence of water on the surface of the earth. Were it not for this property, that is, if ice didn't float, much of our planet's water would be locked up in ice and life would be impossible in its seas, lakes, ponds, and rivers.

Let's examine this in detail to see why. There are many places in the world where the temperature falls below 0°C in winter, often considerably below that. Such cold will of course affect the water in seas, lakes, etc. These bodies of water grow colder and colder and parts of them begin to freeze. If ice didn't behave the way it does (if it didn't float in other words) this ice would sink to the bottom while the warmer bits of water would rise to the surface and be exposed to the air. But the temperature of that air is still below freezing so these will freeze too and sink to the bottom.

This process would continue until there was no liquid water left at all. But this isn't what happens. What happens instead is this: As it gets colder, water grows heavier until it reaches 4°C at which point everything suddenly changes. After this, the water begins to expand and it becomes lighter as the temperature drops. As a result, the 4°C water remains on the bottom, the 3°C water above it, the 2°C water above that and so on. Only at the surface does the temperature of the water actually reach 0°C and there it freezes. But only the surface has frozen: the 4°C layer of water beneath the ice remains liquid and that is enough for underwater creatures and plants to continue to live.

(We should note here that the fifth property of water–the low thermal conductivity of ice and snow–is also crucial in this process. Because they are such poor conductors of heat, the layers of ice and snow keep the heat in the water below from escaping into the atmosphere. As a result of

Unlike all other liquids, water expands when it freezes. Because of this, ice floats in water.

Because water freezes from the top down, the world's oceans remain liquid even though there may be layers of ice on the surface. If water didn't have this "extraordinary" property, nearly all the world's water would be locked up in ice and life in the sea would be impossible.

all this, even if the air temperature falls to –50°C, the layer of sea ice will never be more than a meter or two thick and there will be many fractures in it. Creatures such as seals and penguins that dwell in polar regions can take advantage of this to reach the water beneath the ice.)

Again let us recall what would happen if water didn't behave this way and acted "normally" instead. Suppose water continued to become denser the lower its temperature became like all other liquids and ice sank to the bottom. What then?

Well in that case, the freezing process in the oceans and seas would start from the bottom and continue all the way to the top because there would be no layer of ice on the surface to prevent the remaining heat from escaping. In other words, most of earth's lakes, seas, and oceans would become solid ice with a layer of water perhaps a few meters deep on top of it. Even when the air temperature increased, the ice at the bottom would never melt completely. In the seas of such a world, no life could exist and in an ecological system with dead seas, life on land would also be impossible. In other words, if water didn't "misbehave" and acted normally, our planet would be a dead world.

Why doesn't water act normally? Why does it suddenly begin to expand at 4°C after having contracted the way it should?

That is a question that nobody has ever been able to answer.

Sweat and Cool off

The second and third properties of water mentioned above–high latent heat and thermal capacity greater than other liquids–are also very important for us. These two properties are the keys to an important bodily function whose value we rarely give a thought to. That function is sweating.

Indeed, what good is sweating?

To explain this, we have to take give you a bit of background first. All mammals have bodily temperatures that are fairly close to one another. Although there is some variation, it is not much and mammalian body temperatures range between 35-40°C. In human beings it is about 37°C under normal conditions. This is a very critical temperature and absolutely has to be kept constant. If your body's temperature were to fall just a few degrees, many of its vital functions would fail. If it rises even a few fractions of a degree, as it does when we become ill, the effects can be devastating. A sustained bodily temperature over 40°C is likely to bring on death.

In short, our bodily temperature has a very critical equilibrium in which there is very little room for variation.

However our body has a serious problem here: it is active all the time. All the physical movements, even those of machines, require the production of energy to make them happen. But whenever energy is produced, heat is always generated as a by-product. You can easily see this for yourself. Put this book aside and go take a ten-kilometer run in the blazing sun and see how hot your body gets.

But in fact, if you think about it you'll realize that you didn't get nearly as hot as you should have done...

The thermal properties of water enable us to discharge excessive heat from our body through sweating.

The unit of heat is the calorie. A normal person running 10 kilometers in one hour will generate about 1,000 calories of heat. That heat has to be discharged from the body. If it weren't, you'd collapse into coma before you finished the first kilometer.

That danger however is precluded by the second two properties that water has.

The first of these is the thermal capacity of water. What this means is that in order to increase the temperature of water, a great deal of heat is required. Water makes up about 70% of our body but because of its thermal capacity, that water doesn't get hot very fast. Imagine an action that generates a 10°C increase in bodily heat. If we had alcohol instead of water in our bodies, the same action would lead to a 20°C increase and for other substances with lower thermal capacities the situation would be even worse: increases of 50°C for salt, 100°C for iron, and 300°C for lead. The high thermal capacity of water is what prevents such enormous changes in heat from taking place.

But even an increase of 10°C is would be fatal as we mentioned above. To forestall that, the second property of water–its high latent heat–comes into play.

To keep itself cool in the face of the heat that is being generated, the body employs the sweating mechanism. When we sweat, water spreads over the surface of the skin and quickly evaporates. But because water's latent heat is so great, that evaporation requires large amounts of heat. The heat, of course, is withdrawn from the body and thus we are kept cool. This cooling process is so effective that it can sometimes cause us to experience a chill even when the weather is rather warm.

Because of this, someone who has run ten kilometers will reduce his body temperature by 6°C as a result of the evaporation of just a liter's worth of water. The more energy he expends, the more his body temperature will increase but, at the same time, the more he will sweat and thus cool off. Among the factors that make this magnificent thermostat system of the body possible, foremost are the thermal properties of water. No other liquid would provide for sweating as efficiently as water does. If alcohol were present instead of water for example, the reduction in heat would be only 2.2°C; even in the case of ammonia, it would be only 3.6°C.

There is another important aspect of this matter. If the heat released within the body were not conveyed to the surface, that is to the skin, neither the two properties of water nor the process of sweating would be of any use. Thus the structure of the body must also be highly conductive of heat. It is at this point that another vital property of water appears comes into play: unlike all other known liquids, water has a very high capacity for thermal conductivity, that is, the ability to conduct heat. For this reason, the body conveys the heat generated inside it to the skin. (The blood vessels near the skin expand to achieve this and this is why we become flushed when we're overheated.) If water's thermal conductivity were less by a factor of two or three, the rate of conveyance of heat to the skin would be much slower and this would make it impossible for complex life forms like mammals to live.

What all this shows is that three very different thermal properties of water work together to serve a common purpose: cooling off the bodies of complex life forms such as human beings. Water is a liquid specially designed for this task.

A Temperate World

The five different thermal properties of water mentioned in Henderson's book *The Fitness of Environment* also play a key role in bringing about the mild and balanced climate that Earth has.

Water's greater latent heat and thermal capacity as compared with other liquids are the reasons that bodies of water heat up and cool off more slowly than does the land. On land, the difference in temperature between the hottest and coldest places can reach as high as 140°C; at sea, that difference varies at most between 15-20°C. The same situation exists in the difference between daytime and nighttime temperatures: in arid environments on land, the difference in temperature can be as much as 20-30°C; at sea, this is never more than a few degrees. And not only the seas are affected in this way: the water vapor in the atmosphere is also a big balancing agent. One result of this is that in desert regions where there is very little water vapor present, the difference between daytime and nighttime temperatures is extreme while in regions where a maritime climate prevails, the difference is much less.

Thanks to these unique thermal properties of water, the temperature differences between summer and winter or between night and day remain constantly within limits such that human beings and other living things can survive. If the surface of our world had less water than it does land, the temperature differences between night and day would have been much greater, large tracts of land would have been desert, and life might have been impossible or, at the very least, much more difficult. Similarly, if the thermal properties of water had been different from what they are, the result would have been a planet quite unsuitable for life.

Having examined all these thermal properties of water, Henderson concludes:

> To sum up, this property appears to possess a threefold importance. First, it operates powerfully to equalise and to moderate the temperature of the earth; secondly, it makes possible very effective regulation of the temperature of the living organism; and thirdly it favours the meteorological cycle. All of these effects are true maxima, for no other substance can in this respect compare with water. [79]

The huge volume of water in the earth's seas keeps the planet's temperature in balance. For this reason, the differences between daytime and nighttime temperatures are quite small in regions near the sea, especially along seacoasts. In desert regions far from the sea, the difference between daytime and nighttime temperatures can be as high as 40°C.

High Surface Tension

The properties of water that we have considered till now are thermal: that is, they are its heat-related properties. Water also has a number of physical properties which, as it turns out, are also extraordinarily fit for life.

One of these is water's surface tension, which is extremely high. "Surface tension" is defined as a behavior of the free surface of a liquid to act like an elastic skin under tension. It is caused by attractive forces between the molecules in the surface of the liquid.

The best examples of the effects of surface tension are to be seen in water. Indeed, water's surface tension is so high that some odd physical phenomena take place as a result. A cup can hold a water mass which is slightly higher than its own height without spilling out. A metal needle carefully placed on a motionless watery surface will float.

The surface tension of water is much higher than that of any other known liquid. Some of the biological consequences of this are crucial and this is particularly evident in the case of plants.

Have you ever wondered how plants are able to convey water from the depths of the soil many meters into the air without pumps, muscles, or the like? The answer to this puzzle is surface tension. The channels in the roots and stems of plants are designed to take advantage of water's high surface tension. These channels grow thinner the higher they reach and quite literally cause water to "creep up" on its own.

What makes this excellent design possible is the high surface tension of water. If water's surface tension were as low as it is in most other liquids, it would be physiologically impossible for large plants such as trees to live on dry land.

Another important consequence of water's high surface tension is the fragmentation of rock. Because its surface tension is so high, water is able to penetrate into the deepest recesses of rock through the tiniest of cracks where it freezes when the temperature drops below zero. Water, as we have seen, is unusual in that it expands when it freezes. This expan-

Plants are designed to take advantage of water's high surface tension, thanks to which, water can be made to rise many meters up into even the highest leaves of a forest canopy.

sion exerts interior forces upon rock that causes it eventually to break up. This process is vitally important because it releases the minerals trapped in rock into the environment and also contributes to the formation of soil.

The Chemical Properties of Water

In addition to its physical properties, the chemical properties of water are also extraordinarily fit for life. Foremost among these properties is that it is an excellent solvent: nearly all chemical substances are capable of being dissolved in water.

A very important consequence of this is that useful minerals and similar substances that are locked up in the land get dissolved in water and transported to the sea by rivers. It is estimated that five billion tons of such matter are carried into the sea every year. These substances are vital for sea-life.

Water also accelerates (catalyzes) nearly all known chemical reactions. Another important chemical property of water is that its chemical reactivity is at an ideal level. Water is neither too reactive and thus potentially destructive (as sulfuric acid for example) nor is it too inert (like argon which takes part in no chemical reactions). To quote Michael Denton: "It seems that, like all other properties, the reactivity of water is ideally fit for both its biological and its geological role."[80]

Additional details concerning the fitness of the chemical properties of water for life are constantly being revealed as researchers investigate the matter more. Harold Morowitz, a biophysics professor from the University of Yale, makes this comment:

> The past few years have witnessed the developing study of a newly understood property of water (i.e., proton conductance) that appears to be almost unique to that substance, is a key element in biological-energy transfer, and was almost certainly of importance to the origin of life. The more we learn the more impressed some of us become with nature's fitness in a very precise sense...[81]

Water's Ideal Viscosity

Whenever we think of a liquid, the image that forms in our minds is that of a substance that is extremely fluid. In actual fact, different liquids have highly differing degrees of viscosity: the viscosities of tar, glycerin, olive oil, and sulfuric acid for example vary considerably. And when we compare such liquids with water, the difference becomes even more pronounced. Water is 10 million times more fluid than tar, 1,000 times more so than glycerin, 100 times more than olive oil, and 25 times more than sulfuric acid.

As this quick comparison should indicate, water has a very low degree of viscosity. Indeed, if we discount a few substances such as ether and liquid hydrogen, water appears to have a viscosity that is less than anything except gases.

Does water's low viscosity have any importance for us? Would things be different if this vital liquid were a little more or a little less vis-

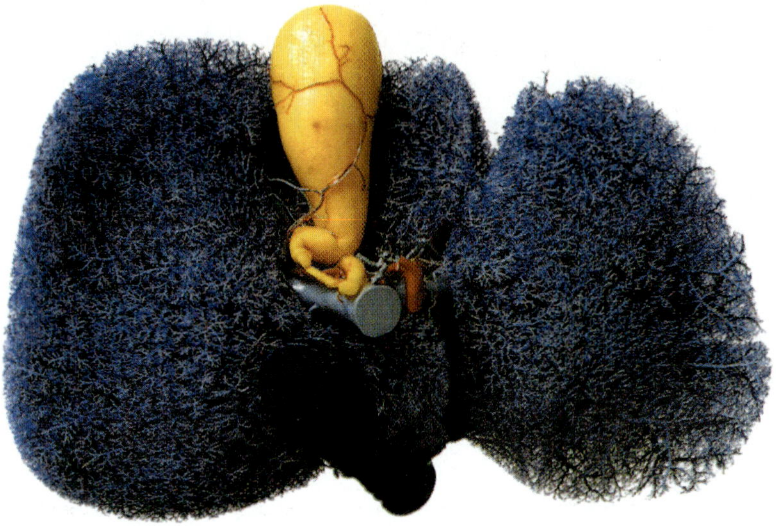

Water's low viscosity is vitally important to us. If water were only slightly more viscous, it would be impossible for blood to be transported through the body's capillary system. For example the complex system of veins of our body's liver (shown at the left) would never have been able to exist.

cous? Michael Denton answers that question for us:

> The fitness of water would in all probability be less if its viscosity were much lower. The structures of living systems would be subject to far more violent movements under shearing forces if the viscosity were as low as liquid hydrogen...If the viscosity of water was much lower, delicate structures would be easily disrupted... and water would be incapable of supporting any permanent intricate microscopic structures. The delicate molecular architecture of the cell would probably not survive.
>
> If the viscosity was higher, the controlled movement of large macromolecules and particularly structures such as mitochondria and small organelles would be impossible, as would processes like cell division. All the vital activities of the cell would be effectively frozen, and cellular life of any sort remotely resembling that with which we are familiar would be impossible. The development of higher organisms, which is critically dependent on the ability of cells to move and crawl around during embryogenesis, would certainly be impossible if the viscosity of water was even slightly greater than it is.[82]

Water's low viscosity is essential not only for cellular motion but also for the circulatory system.

All living creatures with a body size of more than a quarter of a mil-

Water's low viscosity is essential for all living things, even plants. The tiny veins in the leaf seen at the left are able to transport water because it is so fluid.

limeter have a centralized circulatory system. The reason is that beyond that size, it is not possible for nutriments and oxygen to be diffused throughout the organism. That is, they can no longer be taken directly into the cell nor can their by-products be discharged. There are many cells in an organism's body and thus it is necessary for the oxygen and energy taken into the body to be distributed (pumped) to them through "ducts" of some sort; similarly, other channels are necessary to carry away the waste. These "ducts" are the veins and arteries of the circulatory system. The heart is the pump that keeps this system moving while the substance carried through the "ducts" is the liquid we call "blood", which is mostly water. (95% of blood plasma–the material remaining after blood cells, proteins, and hormones have been removed, is water.)

This is why the viscosity of water is so important for the efficient functioning of the circulatory system. If water had the viscosity of tar for example, certainly no organic heart could pump it. If water had the viscosity even of olive oil, which is a hundred million times less viscous than tar, the heart might be able to pump it, but it would be extremely difficult and blood would never be able to reach all the billions of capillaries that wend their ways through our bodies.

Let's take a closer look at those capillaries. Their purpose is to carry the oxygen, nourishment, hormones, etc that are necessary for life to every cell everywhere in the body. If a cell is more than 50 microns (a micron is a thousandth of a millimeter) away from a capillary it cannot take advantage of the capillary's "services". Cells more than 50 microns from a capillary will starve to death.

This is why the human body was so created that the capillaries form a network that pervades it completely. A normal human body has about 5 billion capillaries whose total length, if stretched out, is about 950 kilometers. In some mammals, there are as many as 3,000 capillaries in a single square centimeter of muscle tissue. If you were to gather ten thousand of the tiniest capillaries in the human body together, the resulting bundle might be as thick as the lead of a pencil. The diameters of these capillaries varies between 3-5 microns: that's three to five thousandths of a millimeter.

If blood is going to penetrate passages that narrow without blocking them or slowing down, it certainly needs to be fluid and, thanks to water's low viscosity, it is. According to Michael Denton, if water's viscosity were just a bit more than what it is, the blood circulatory system would be completely useless:

> A capillary system will work only if the fluid being pumped through its constituent tubes has a very low viscosity. A low viscosity is essential because flow is inversely proportional to the viscosity... From this it is easy to see that **if the viscosity of water had a value only a few times greater than it is**, pumping blood through a capillary bed would require enormous pressure and **almost any sort of circulatory system would be unworkable**... If the viscosity of water had been slightly greater and the smallest functional capillaries had been 10 microns in diameter instead of 3, then the capillaries would have to occupy virtually all of the muscle tissue to provide an effective supply of oxygen and glucose. Obviously the design of macroscopic life forms would be impossible or enormously constrained... It seems, then, the viscosity of water must be very close to what it is if water is to be a fit medium for life.[83]

In other words, like all its other properties, the viscosity of water is also "tailor-made" for life. Looking at the viscosities of different liquids, we see that they differ by factors of many billions. Among all those billions there is one liquid whose viscosity has been created to be exactly what it needs to be: water.

Conclusion

Everything that we have seen in this chapter since its beginning shows us that the thermal, physical, chemical, and viscosity properties of water are exactly what they must be in order for life to exist. Water is so perfectly designed for life that, in some cases, the very laws of nature are suspended to make it so. The best example of this is the unexpected and inexplicable expansion that takes place in water's volume when its temperature falls below 4°C: if that didn't happen ice wouldn't float, the seas would freeze all but solid, and life would be impossible.

Water is "just right" for life to a degree that cannot be compared with any other liquid. The larger part of this planet, a world whose other attributes (temperature, light, electromagnetic spectrum, atmosphere, surface, etc) are all suitable for life, has been filled with just the right amount of water necessary for life. It should be obvious that this cannot all be accidental and that there must instead be intentional design.

To put it another way, all the physical and chemical properties of water show us that it is created especially for life. The earth, purposefully created for mankind to live in, was brought to life with this water that was specially created to form the basis of human life. In water, Allah has given us life and with it He causes the food by which we are nourished to spring from the soil.

But the most important aspect of all this is that this truth, which has been discovered by modern science, was revealed in the Qur'an, bestowed upon humanity as a guide fourteen centuries ago. Concerning water and mankind, Allah's word is revealed in the Qur'an thus:

> **It is He who sends down water from the sky. From it you drink and from it come the shrubs among which you graze your herds. And by it He makes crops grow for you and olives and dates and grapes and fruit of every kind. There is certainly a Sign in that for people who reflect. (Surat an-Nahl: 10-11)**

CHAPTER VIII

THE SPECIALLY-DESIGNED ELEMENTS OF LIFE

There is a mind and purpose behind the universe. There are hints of that divine presence in how abstract mathematics can penetrate the universe's secrets, which suggests that a rational mind created the world. Nature is fined tuned to allow life and consciousness to emerge.

John Polkinghorne, British Physicist[84]

Up to this point we have been examining how all the physical balances of the universe in which we live have been specially designed so that we can live. We have seen how the general structure of this universe, the location of Earth in it, and factors such as air, light, and water have been precisely designed to have exactly the attributes we require. In addition to all this however, we also need to take a look at the elements that make up our bodies. These chemical elements, the building-blocks from which our hands, eyes, hair, and organs as well as all the living things–plants and animals–that are our sources of food have been specially designed to serve the exact purposes that they do.

The physicist Robert E. D. Clark refers to the special and superior design in the building-blocks of life when he says: "As if the Creator has given us a kit of prefabricated parts ready made for the work in hand."[85]

The most important of these building-blocks is carbon.

The Design in Carbon

In previous chapters we described the extraordinary process by which carbon, the element that occupies the sixth position in the periodic table, was produced in the hearts of the huge stars called "red giants". We also saw how, having discovered this wonderful process, Fred Hoyle was moved to say that "the laws of nuclear physics have been deliberately designed with regard to the consequences they produce inside the stars."[86]

When we examine carbon more closely, we can see that not just the physical formation of this element but also its chemical properties were deliberately arranged to be what they are.

Pure carbon occurs naturally in two forms: graphite and diamonds. Carbon however also enters into compounds with many other elements and the result is many different kinds of substances. In particular, the incredibly varied range of organic materials of life–the membrane of a cell and the bark of a tree, the lens of an eye and the horn of a deer, the white of an egg and the poison of a snake–are all made up of carbon-based com-

One natural form of pure carbon is graphite. This element however is able to form an extraordinary range of different substances when it combines with the atoms of other elements. The main structure of the human body is the result of the different chemical bonds that carbon is capable of entering into.

pounds. Carbon, combined with hydrogen, oxygen, and nitrogen in many different quantities and geometric arrangements, results in a vast assortment of materials with vastly different properties.

Some carbon compounds' molecules consist of just a few atoms; others contain thousands or even millions. Furthermore, no other element is as versatile as carbon is in forming molecules with such durability and stability. To quote David Burnie in his book *Life*:

> Carbon is a very unusual element. Without the presence of carbon and its unusual properties, it is unlikely that there would be life on Earth.[87]

Concerning carbon, the British chemist Nevil Sidgwick writes in *Chemical Elements and their Compounds:*

> Carbon is unique among the elements in the number and variety of the compounds which it can form. Over a quarter of a million have already been isolated and described, but this gives a very imperfect idea of its powers, since it is the basis of all forms of living matter.[88]

For reasons of both physics and chemistry, it is impossible for life to be based on any element other than carbon. At one time, silicon was proposed as another element on which life might be based. We now know however that this conjecture is impossible. To quote Sidgwick again:

> We know enough now to be sure that the idea of a world in which silicon should take the place of carbon as the basis of life is impossible…[89]

Covalent Bonds

The chemical bonds that carbon enters into when forming organic compounds are called "covalent bonds". A covalent bond is said to occur when two atoms share their electrons.

The electrons of an atom occupy specific orbital shells that are centered around the nucleus. The orbit closest to the nucleus can be occupied by no more than two electrons. In the next orbit a maximum of eight electrons is possible. In the third orbit, there can be up to eighteen. The number of electrons continues to increase with the addition of more orbits. Now an interesting aspect of this scheme is that atoms seem to "want" to complete the number of electrons in their orbital shells. Oxygen, for example, has six electrons in its second (and outermost) orbit, and this makes it "eager" to enter into combinations with other atoms that will supply the two more electrons that are needed to increase this number to eight. (Why atoms behave this way is a question that is unanswered. But it's a good thing they do: because if they didn't, life wouldn't be possible.)

Covalent bonds are the result of this tendency of atoms to complete their orbital shells. Two or more atoms can often make up the shortfall in their orbits by sharing electrons with one another. A good example is the water molecule (H_2O), whose building-blocks (two hydrogen atoms and one oxygen atom) form a covalent bond. In this compound, oxygen completes the number of electrons in its second orbit to eight by sharing the two electrons (one each) in the orbital shells of the two hydrogen atoms; in the same way, the hydrogen atoms each "borrow" one electron from oxygen to complete their own shells.

Carbon is very good at forming covalent bonds with other atoms (including carbon atoms) from which an enormous number of different compounds can be made. One of the simplest of these compounds is methane: a common gas that is formed from the covalent bonding of four hydrogen atoms and one carbon atom. With only six electrons, carbon's outer orbital shell is short of the eight that it needs by four, rather than two as is the case with oxygen, and for this reason, four hydrogen atoms are needed to complete it.

We said that carbon was especially versatile in forming bonds with other atoms and this versatility makes an enormous number of different compounds possible. The class of compounds formed exclusively from carbon and hydrogen are called **"hydrocarbons"**. This is a huge family of compounds that includes natural gas, liquid petroleum, kerosene, and lubricating oils. Hydrocarbons like ethylene and propylene are the "bedrock" on which the modern petrochemical industry has been erected. Hydrocarbons like benzene, toluene, and turpentine are familiar to anyone who's worked with paints. The naphthalene that protects our clothes from moths is another hydrocarbon. With the addition of chlorine in their composition, some hydrocarbons become anesthetics; with the addition of fluorine, we have Freon, a gas that is widely used in refrigeration.

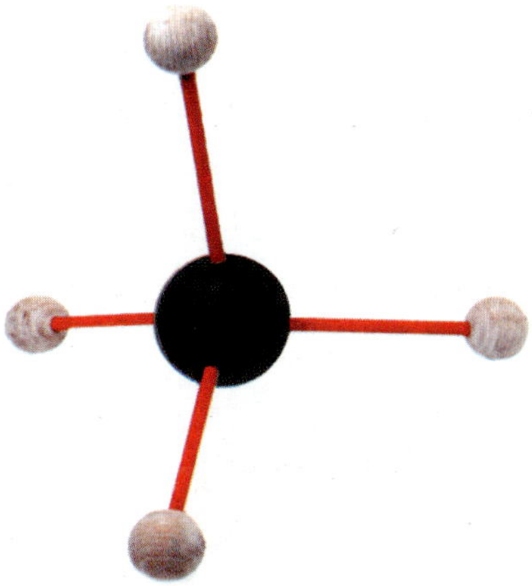

The structure of methane: Four hydrogen atoms share one electron each with a single carbon atom.

There is another important class of compounds in which carbon, hydrogen, and oxygen form covalent bonds with one another. In this family we find alcohols like ethanol and propanol, ketones, aldehydes, and fatty acids among many, many other substances. Another group of compounds composed of carbon, hydrogen, and oxygen are sugars, including glucose and fructose.

The cellulose that makes up the skeleton of wood and the raw material for paper is a carbohydrate. So is vinegar. So is beeswax and formic acid. Each one of the incredibly rich panoply of substances and materials that occur naturally in our world is "nothing more" than a different arrangement of carbon, hydrogen, and oxygen linked together by covalent bonds.

Olive oil, meat, and brown sugar: Everything we eat is made up of different arrangements of hydrogen, oxygen, and carbon with the addition of other atoms such as nitrogen.

When carbon, hydrogen, oxygen, and nitrogen form such bonds, the result is a class of molecules that is the foundation and structure of life itself: the amino acids that make up proteins. The nucleotides that make up DNA are also molecules formed from carbon, hydrogen, oxygen, and nitrogen.

In short, the covalent bonds that the carbon atom is capable of entering into are vital for the existence of life. Were hydrogen, carbon, nitrogen, and oxygen not so "eager" to share electrons with one another, life would indeed be impossible.

The thing that makes it possible for carbon to form these bonds is a property that chemists call "metastability", the characteristic of having only a slight margin of stability. The biochemist J. B. S. Haldane describes metastability thus:

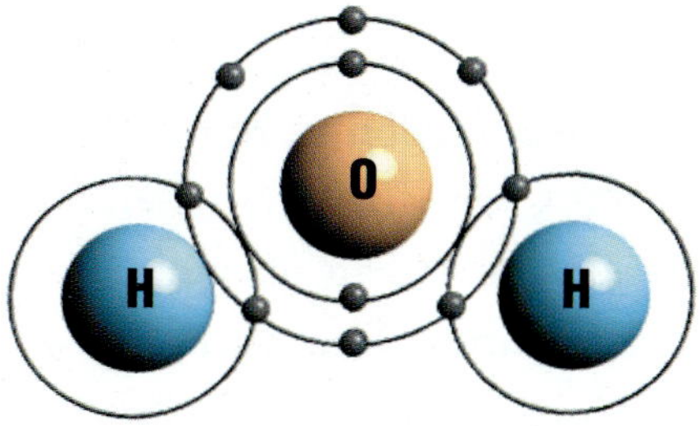

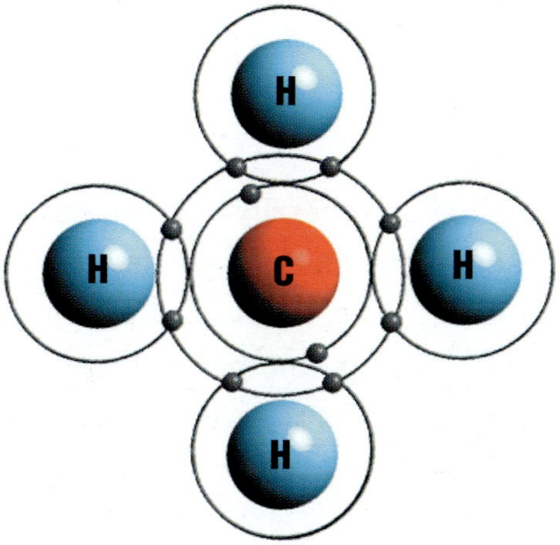

WATER AND METHANE: TWO DIFFERENT EXAMPLES OF COVALENT BONDS

In the water molecule (above), there is a covalent bond between the two hydrogen atoms and the one oxygen atom. In the methane molecule below, four hydrogen atoms form covalent bonds with a single carbon atom.

> A metastable molecule means one that can liberate free energy by a transformation, but is stable enough to last a long time unless it is activated by heat, radiation, or union with a catalyst.[90]

What this somewhat technical definition means is that carbon has a rather unique structure, thanks to which, it is quite easy for it to enter into covalent bonds under normal conditions.

But it is precisely here that the situation starts to become curious because **carbon is metastable only within a very narrow range of temperatures.** Specifically, carbon compounds become very unstable when the temperature goes over 100°C.

This fact is so commonplace in our everyday lives that most of us take it for granted. When we cook meat for example, what we're really doing is changing the structure of its carbon compounds. But there's a point here that we should note: The cooked meat has become completely "dead"; that is, its chemical structure is different from what it had when it was part of a living organism. Indeed **most carbon compounds become "denatured" at temperatures above 100°C**: the majority of vitamins for example simply fall apart at that temperature; sugars also undergo structural changes and lose some of their nutritional value; and at around 150°C, carbon compounds will start to burn.

In other words, if carbon atoms are to enter into covalent bonds with other atoms and if the resulting compounds are to remain stable, the ambient temperature must not go over 100°C. The lower boundary on the other hand is around 0°C: if the temperature drops too much below that, organic biochemistry becomes impossible.

In the case of other compounds, this is generally not the situation. Most inorganic compounds are not metastable; that is, their stability is not greatly affected by changes in temperature. To see this let's do an experiment. Stick a piece of meat on the end of a long, thin piece of metal such as iron and heat the two together over a fire. As the temperature grows hotter, the meat will darken and eventually burn long before much of anything happens to the metal. The same thing would be true if you substituted stone or glass for metal. You would have to increase the heat by many hundreds of degrees before the structures of such materials began to change.

By now you certainly will have spotted the similarity between the temperature range that is necessary for carbon compounds' covalent bonds to be established and remain stable and the range of temperatures that prevails on our planet. As we have said elsewhere, in the whole universe, temperatures range from the millions of degrees in the hearts of stars to absolute zero (-273.15°C). But Earth, having been created for humanity to live in, possesses the narrow temperature range essential for the formation of the carbon compounds that are the building-blocks of life.

But the curious "coincidences" do not end here. The same temperature interval is the only one in which water remains liquid. As we saw in the earlier chapter, liquid water is one of the basic requirements of life and, in order to remain liquid, it requires precisely the same temperatures that carbon compounds need to form and be stable. There is no physical or natural "law" dictating that this should be so and under the circumstances, this situation is evidence that the physical properties of water and carbon and the conditions of the planet Earth were created so as to be in harmony with one another.

Weak Bonds

Covalent bonds are not the only type of chemical bonding that keeps the compounds of life stable. There is another and different category of bond known as "weak bonds".

Such bonds are about twenty times weaker than covalent bonds, hence their name; but they are no less crucial to the processes of organic chemistry. It is thanks to this weak bonding that the proteins that make up the building-blocks of living things are able to maintain their complex and vitally important three-dimensional structures.

To explain this, we have to talk briefly about the structure of proteins. Proteins are usually referred to as a "chain" of amino acids. While this metaphor is essentially correct, it is also incomplete. It's incomplete because for most people a "chain of amino acids" conjures up the mental image of something like a string of pearls whereas the amino acids that make up proteins have a three-dimensional structure more like a tree with leafy branches.

Covalent bonds are what hold the atoms of amino acids together. Weak bonds are what maintain the essential three-dimensional structure of those acids. No proteins could exist without these weak bonds. And of course without proteins, there could be no life.

Now the interesting part of this business is that the temperature range in which weak bonds are able to perform their function is the same as the one prevailing on Earth. This is rather odd because the physical and chemical natures of covalent bonds versus weak bonds are entirely different things and independent of one another. In other words, there's no

Covalent bonds: The atoms are strongly linked to one another

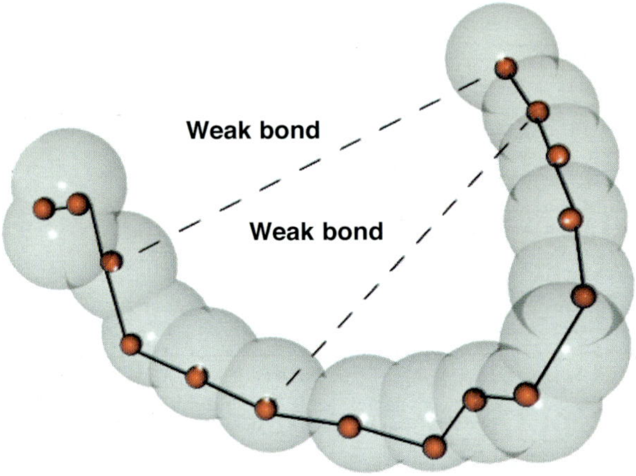

Weak bonds: An organic compound is maintained in a unique three-dimensional form by means of weak (non-covalent) bonds (the broken lines).

intrinsic reason why they should both require the same temperature range. And yet they do: Both types of bonds can only be formed and remain stable within this narrow temperature range. And if they did not–if covalent bonds required a range of temperatures wildly different from that of weak bonds, say–then it would be impossible to construct the complex three-dimensional structures that proteins require.

Everything that we have seen concerning the extraordinary chemical properties of the carbon atom shows that there is an enormous harmony existing among this element that is the fundamental building-block of life, the water that is also vital for life, and the planet Earth that is the shelter for that life. In *Nature's Destiny*, Michael Denton underscores this fitness when he says:

> Out of the enormous range of temperatures in the cosmos, there is only one tiny temperature band in which we have (1) liquid water, (2) a great plenitude of metastable organic compounds, and (3) weak bonds for stabilizing the 3-D forms of complex molecules.[91]

Among all the heavenly bodies that have ever been observed, this "tiny temperature band" exists only on Earth. Moreover it is only on Earth that the two fundamental building-blocks of life—carbon and water–are to be found in such generous supply.

What all this indicates is that the carbon atom and its extraordinary properties were specially designed for life and that our planet was specially created to be a home for carbon-based life-forms

The Design in Oxygen

We have seen how carbon is the most important building-block of living organisms and how it was specially designed so as to fulfill that function. The existence of all carbon-based life-forms however is contingent upon a second imperative: energy. Energy is an indispensable requirement for life.

Green plants get their energy from the sun through the process of photosynthesis. For the rest of the living creatures of Earth–and that includes us–the only source of energy is a process called "oxidation"–a fancy word for "burning". The energy of oxygen-breathing organisms is derived from burning the nourishment that they get from plants and animals. As you may guess from the term "oxidation", this burning is a chemical reaction in which substances are oxidized–that is, they are combined with oxygen. This is why oxygen is as vitally important to life as are carbon and hydrogen.

A generalized formula for "burning" (oxidation) looks like this:

Carbon compound + oxygen > water + carbon dioxide + energy

What this means is that when carbon compounds and oxygen are combined (under the proper conditions of course) a reaction takes place that generates water and carbon dioxide and releases a considerable amount of energy. This reaction takes place most readily in hydrocarbons (compounds of hydrogen and carbon). Glucose (a sugar and also a hydrocarbon) is what is constantly being burned in your body to keep it supplied with energy.

Now as it happens, the elements of hydrogen and carbon that make up hydrocarbons are the ones most suitable for oxidation to take place. Among all other atoms, hydrogen combines with oxygen the most readily and releases the most energy in the process. If you need a fuel to burn in oxygen, you can't do better than hydrogen. From the standpoint of its value as a fuel, carbon ranks third after hydrogen and boron. In *The Fitness of the Environment*, Lawrence Henderson comments on the **extraordinary fitness** that is involved here:

> The very chemical changes, which for so many other reasons seem to be best fitted to become the processes of physiology, turn out to be the very ones which can divert the greatest flood of energy into the stream of life.[92]

The Design in Fire (Or Why You Don't Just Burst Into Flame)

As we've just seen, the fundamental reaction that releases the energy necessary for the survival of oxygen-breathing organisms is the oxidation of hydrocarbons. But this simple fact raises a troubling question: If our bodies are made up essentially of hydrocarbons, why aren't they also oxidized? Putting it another way, why don't we just go up in flame, like a match that's been struck?

Our bodies are constantly in contact with the oxygen of the air and yet they don't oxidize: they don't catch fire. Why not?

The reason for this seeming paradox is that, under normal conditions of temperature and pressure, the molecular (O_2) form of oxygen has a substantial degree of inertness or "nobility". (In the sense that chemists use the term, "nobility" is the reluctance (or inability) of a substance to enter into chemical reactions with other substances.) But this raises another questions: If molecular oxygen is so "noble" as to avoid incinerating us, how is this same molecule made to enter into chemical reactions inside our bodies?

The answer to this question, which perplexed chemists as early as the mid 19th century, did not become known until the second half of the 20th century, when biochemical researchers discovered the existence of enzymes in the human body whose only function was to force the O_2 in the atmosphere to enter into chemical reactions. As a result of a series of extremely complex steps, these enzymes utilize atoms of iron and copper in our bodies as catalysts. A catalyst is a substance that initiates a chemical reaction and allows it to proceed under different conditions (such as lower temperature etc) than would otherwise be possible.[93]

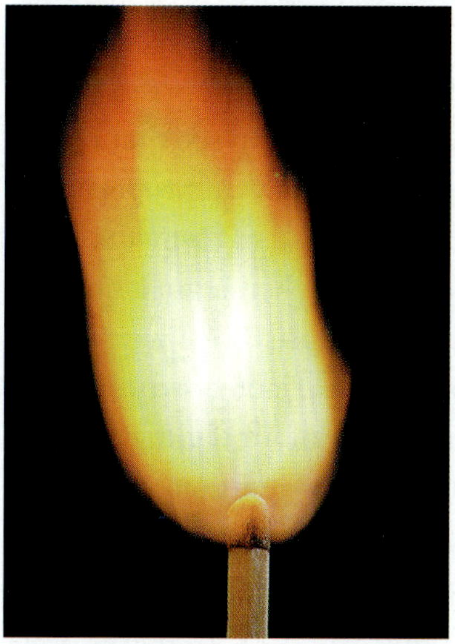

In other words, there is a very interesting situation here: Oxygen is what supports oxidation and combustion and normally one would expect it to burn us up too. To prevent this, the molecular O_2 form of oxygen that exists in the atmosphere has been given a strong element of chemical nobility. That is, it doesn't enter into reactions easily. But, on the other hand, our bodies depend upon the oxidizing property of oxygen for their energy and for that reason, our cells have been fitted out with an extremely complex enzyme system that makes this noble gas extremely reactive.

While we're on the subject we should also point out that this enzyme system is a marvellous example of design that no evolutionary theory holding that life developed as a result of chance events can ever hope to explain.[94]

There is yet another precaution that has been taken to keep our bodies from burning up: what the British chemist Nevil Sidgwick calls the "characteristic inertness of carbon".[95] What this means is that carbon is not too much in a hurry either to enter into a reaction with oxygen under normal pressures and temperatures. Expressed in the language of chemistry this may all seem rather arcane, but in fact what is being said here is something that anyone who's ever had to light a fireplace full of huge logs or a coal-burning stove in winter or start a stubborn barbecue in summer already knows. In order to get the fire going, you have to take care of a lot of preliminaries (kindling, starter, etc) or else suddenly raise the temperature of the fuel to a very high degree (as with a blowtorch). But once the fuel starts burning, the carbon in it enters into the reaction with oxygen quite rapidly and a great amount of energy is released. This is why it's so hard to get a fire going without another source of heat. But after combustion begins, a great deal of heat is produced and this can cause other carbon compounds nearby to catch fire as well and so the fire spreads.

When we look into this matter more carefully, we can see that **fire itself is a most interesting example of design**. The chemical properties of oxygen and carbon have been so arranged that these two elements enter into a reaction with one another (combustion) only when a great amount of heat is already present. It's a good thing, too because if this weren't the case, life on this planet would be very unpleasant if not downright impossible. If oxygen and carbon were even slightly more willing to react with one another, the spontaneous combustion–self-ignition–of people, trees, and animals would become a commonplace event whenever the weather got a little too warm. Someone walking through a desert for example might suddenly burst into flame at noon when the heat was at its most intense; plants and animals would be exposed to the same risk. Even if life were possible in such a world, it certainly wouldn't be much fun.

On the other hand, if carbon and oxygen were slightly more noble (that is, slightly less reactive) than they are, it would be much more difficult to light a fire in this world than it already is: indeed, it might even be impossible. And without fire, we not only would have been unable to keep ourselves warm: it's quite likely that there would never have been any technological progress on our planet because that progress depends upon the ability to work materials such as metal and without the heat provided by fire, purifying and working metal is all but impossible.

What all this shows is that the chemical properties of carbon and oxygen have been arranged so as to be the most suitable for the needs of mankind. Concerning this, Michael Denton says:

> This curious unreactivity of the carbon and oxygen atoms at ambient temperatures, combined with the enormous energies inherent in their combination once achieved, is of great adaptive significance to life on Earth. It is this curious combination that not only makes available to advanced life forms the vast energies of oxidation in a controlled and orderly manner but has also made possible the controlled use of fire by mankind and allowed the harnessing of the massive energies of combustion for the development of technology.[96]

In other words, both carbon and the oxygen have been created with properties that are the most fit for human life. The properties of these two elements allow us to light a fire and to make use of fire in the most convenient way possible. Furthermore, the world is full of sources of carbon (such as the wood of trees) that are fit for combustion. All this is an indication that fire and the materials to start and sustain it have been specially created to be fit for human life. In the Qur'an, Allah speaks to mankind with these words:

> **He Who produces fire for you from green trees so that you use them to light your fires. (Surah Ya-sin: 80)**

The Ideal Solubility of Oxygen

The utilization of oxygen by the body is highly dependent upon the property of this gas to dissolve in water. The oxygen that enters our lungs when we inhale is immediately dissolved into the blood. The protein called hemoglobin captures these oxygen molecules and carries them to the other cells of the body where, thanks to the special enzyme system described above, the oxygen is used to oxidize carbon compounds called ATP to release their energy.

All complex organisms derive their energy in this way. However the operation of this system is especially dependent upon the solubility of oxygen. If oxygen were not sufficiently soluble, not enough oxygen would enter the bloodstream and cells would not be able to generate the energy they require; if oxygen were too soluble on the other hand, there would be an excess of oxygen in the blood resulting in a condition known as oxygen toxicity.

The difference in the water-solubility of different gases varies by as much as a factor of a million. That is, the most soluble gas is a million times more soluble in water than the least soluble gas is and there are hardly any gases at all whose solubilities are identical. Carbon dioxide is about twenty times more soluble in water than oxygen is for example. Among the vast range of potential solubilities however, the one possessed by oxygen is precisely what it needs to be for it to be fit for human life.

What would happen if the water-solubility rate of oxygen were different: a little more or a little less?

Let us take a look at the first situation. If oxygen were less soluble in water (and thus also in blood) less oxygen would enter the bloodstream and the body's cells would be starved of oxygen. This would make life much more difficult for metabolically active organisms such as human beings. No matter how hard you worked at breathing, you would constantly be faced with the danger of suffocation because not enough oxygen was reaching your body's cells.

If the water-solubility of oxygen were higher on the other hand, you would be confronted by the threat of oxygen toxicity, mentioned briefly above. Oxygen is, in fact, a rather dangerous substance: if an organism gets too much of it, the result can be fatal. Some of the oxygen in the blood enters into a chemical reaction with the blood's water. If the amount of dissolved oxygen becomes too high, the result is the production of highly reactive and damaging by-products. One of the functions of the complex system of blood enzymes is to prevent this from happening. But if the amount of dissolved oxygen becomes too high, the enzymes cannot do their job. As a result, every breath we take would poison us a little bit more leading quickly to death. The chemist Irwin Fridovich comments on this issue:

> All respiring organisms are caught in a cruel trap. The very oxygen which supports their lives is toxic to them and they survive precariously, only by virtue of elaborate defense mechanisms.[97]

What saves us from this trap–from being poisoned by too much oxygen or from being suffocated by not enough of it–is the fact that oxygen's solubility and the body's complex enzymatic system have been carefully designed and created to be what they need to be. To put it more explicitly, Allah has created not only the air we breathe but also the systems that make it possible to use that air in perfect harmony with one another.

The Other Elements

Carbon and oxygen of course are not the only elements that have been deliberately designed to make life possible. Elements like hydrogen and nitrogen, which make up a large part of the bodies of living things, also possess attributes that make life possible. In fact, there appears not to be a single element in the periodic table that does not fulfill some sort of function in support of life.

In the basic periodic table there are ninety-two elements ranging from hydrogen (the lightest) to uranium (the heaviest). (There are of course other elements beyond uranium but these do not occur naturally and have all been created under laboratory conditions. None of them are stable.) Of this ninety-two, twenty-five are directly necessary for life and of those, just eleven–hydrogen, carbon, oxygen, nitrogen, sodium, magnesium, phosphorus, sulfur, chlorine, potassium, and calcium–make up some 99% of the body weight of nearly all living things. The other fourteen elements (vanadium, chromium, manganese, iron, cobalt, nickel, copper, zinc, molybdenum, boron, silicon, selenium, fluorine, and iodine) are present in living organisms only in very small amounts but even these have vitally important functions. Three elements–arsenic, tin, and tungsten–are to be found in some living things where they perform functions that are not completely understood. Three more elements–bromine, strontium, and barium–are known to be present in most organisms, but their functions are still a mystery.[98]

This broad spectrum encompasses atoms from each of the different series of the periodic table, whose elements are grouped according to the attributes of their atoms. What this indicates is that all of the element groups of the periodic table are necessary, in one way or another, for life. In *The Biological Chemistry of the Elements*, J. J. R Frausto da Silva and R. J. P Williams have this to say:

> The biological elements seem to have been selected from practically all groups and subgroups of the periodic table... and this means that practically all kinds of chemical properties are associated with life processes within the limits imposed by environmental constraints.[99]

Even the heavy, radioactive elements at the end of the periodic table have been marshaled in the service of human life. In *Nature's Destiny*, Michael Denton describes in detail the essential role that these radioactive elements, such as uranium, play in the formation of the earth's geological structure. Naturally occurring radioactivity is closely associated with the fact that the earth's core is able to retain its heat. That heat is what keeps the core, which consists of iron and nickel, liquid. This liquid core is the source of the earth's magnetic field which, as we have seen elsewhere, helps shield the planet from dangerous radiation and particles from space while performing other functions as well. Even the inert gases and elements such as the rare-earth metals, none of which seem to be involved in the support of life, are apparently there because of the demands of ensuring that the range of naturally-occurring elements would extend as far as uranium.[100]

In short, it is safe to say that all the elements whose existence we know of serve some function in human life. Not one of them is either superfluous or purposeless. This situation is further evidence that the universe was created by Allah for mankind.

Conclusion

Every physical and chemical property of the universe that we have examined turns out to be exactly what it needs to be in order for life to exist. And yet in this book we have only scratched the surface of the overwhelming evidence of this fact. No matter how deeply you delve the details or broaden the search, this general observation remains true: In every detail of the universe, there is a purpose that serves human life and each detail is perfectly designed, balanced, and harmonized to achieve that purpose.

Certainly this is proof of the existence of a superior creator who brought this universe into being for this purpose. Whatever property of matter we may examine, we behold in it the infinite knowledge, wisdom, and power of Allah, Who created it from nothingness. Every thing bows to His will and that is why each and every thing is in perfect harmony with everything else.

This is the conclusion that 20th-century science has at last reached. And yet, it is only a recognition of a fact that was imparted to mankind in the Qur'an over fourteen centuries ago: Allah has created every detail of the universe to reveal the perfection of His own creation:

> **Blessed be He who has the Kingdom in His Hand! He has power over all things. He who created the seven heavens in layers. You will not find any flaw in the creation of the All-Merciful. Look again-do you see any gaps? Then look again and again. Your sight will return to you dazzled and exhausted. (Surat al-Mulk: 1-4)**

$$dt_{Schw,comobs} = da\,\frac{\partial t(a,\eta)}{\partial a}\bigg|_{\eta=\eta_{comobs}}$$

$$= -t_0\,\frac{\zeta^3}{(1+\zeta^2)^2(\zeta^2-b^2)}\left(\frac{1-\eta_{edge}^2}{1-\eta^2}\right)^{1/2}\frac{a_{max}}{(a_{max}-a)^2}\,da$$

$$\beta(r,t)=\frac{\left[1-\frac{a_{max}}{a}\left(1-\frac{\left(1-\eta_{edge}^2\right)^{3/2}}{\sqrt{1-\eta^2}}\right)\right]^2}{\left(1-\frac{a_{max}}{a}\eta^2\right)\left[1-\frac{a_{max}}{a}\left(1-\frac{\sqrt{1-\eta_{edge}^2}}{\sqrt{1-\eta^2}}\right)\right]^3}$$

$$r=\eta a$$

$$t=\pm\frac{t_0}{1+b^2}\left[\frac{b^3}{1+b^2}\ln\left(\frac{\zeta+b}{\zeta-b}\right)+\frac{\zeta}{1+\zeta^2}+\frac{1+3b^2}{1+b^2}\left(\frac{\pi}{2}-\arctan\zeta\right)\right]$$

$$t_0\equiv\frac{a_{max}}{c\sqrt{1-\eta_{edge}^2}}$$

$$\zeta\equiv\sqrt{\frac{a_{max}}{a_{max}-a}\sqrt{\frac{1-\eta_{edge}^2}{1-\eta^2}}-1}$$

$$b\equiv\frac{\eta_{edge}}{\sqrt{1-\eta_{edge}^2}}$$

$$ds^2=c^2d\tau_{comobs}^2$$
$$=\beta\left(r_{comobs},t_{comobs}\right)dt_{comobs}^2-\alpha\left(r_{comobs},t_{comobs}\right)d$$

$$d\tau_{comobs}=\pm\frac{da}{c}\sqrt{\frac{a}{a_{max}-a}}=d\tau_c$$

$$ds^2=\beta(r,t)c^2dt^2-\alpha(r,t)dr^2-r^2d\Omega^2$$

$$\alpha(r,t)=\frac{1}{1-\frac{a_{max}r^2}{a^3}}$$

CONCLUSION

AN APPEAL TO REASON

The belief that our wondrous universe could have evolved by blind chance is crazy. And I do not at all mean crazy in the sense of a slangy invective but rather in the technical meaning of psychotic. Indeed such a view has much in common with certain aspects of schizophrenic thinking.

Karl Stern, University of Montreal Psychiatrist[101]

At the beginning of this book we made mention of something called the anthropic principle and said that it was gaining widespread acceptance in the scientific world. As we pointed out then, the anthropic principle holds that the universe is not a purposeless, pointless, or random conglomeration of matter and that, on the contrary, it was carefully and deliberately designed to serve as a home for human life.

Since then we have seen a host of evidence demonstrating that the anthropic principle is indeed a fact: evidence ranging from the speed at which the Big Bang was propagated to the physical balances of atoms, from the relative strengths of the four fundamental forces to the alchemy of stars, from the mysteries of the dimensions of space to the layout of the solar system. And everywhere we've looked we have seen an extraordinarily precise arrangement in the structure of the universe. We saw how the structuring and dimensioning of the world in which we live and even of its atmosphere are exactly what they need to be. We witnessed how the light sent to us by the sun, the water we drink, and the atoms that make up our bodies and the air that we inhale constantly into our lungs are all amazingly fit for life.

In short, any time we observe anything in the universe we encounter an extraordinary design whose purpose is to nurture human life. To deny the reality of this design is, as the psychiatrist Karl Stern put it, to overstep the bounds of reason.

The implications of this design are also obvious. The design concealed within every detail of the universe is most certainly proof of the existence of a Creator who is in control of every detail and whose power and wisdom are infinite. As the Big Bang theory has revealed, this same Creator created the universe from nothingness.

This conclusion that has been reached by modern science is a fact imparted to us in the Qur'an: Allah created the universe from nothingness and gave it order:

Your Lord is Allah, Who created the heavens and the earth in six days and then settled Himself firmly on the Throne. He covers the day with the

night, each pursuing the other urgently; and the sun and moon and stars are subservient to His command. Both creation and command belong to Him. Blessed be Allah, the Lord of all worlds. (Surat al-A'raf: 54)

Unsurprisingly, the discovery of this truth by science upset quite a few scientists and it continues to do so. These are scientists who equate science with materialism; they are people who are convinced that science and religion can never get along and that being "scientific" is synonymous with being an atheist. They have been trained to believe that the universe and all the life in it can be explained as the product of chance events that are completely devoid of any intention or design. When such people encounter the obvious fact of creation, their great dismay and confusion are natural.

In order to understand the consternation of materialists, we need to take a brief look at the question of the origin of life.

The Origin of Life

The origin of life, which is to say, the question of how the first living things came into being on Earth, is one of the biggest dilemmas confronting materialists in the last century and a half. Why should that be so? It's because even a single living cell, the smallest unit of life, is incomparably more complex than even the greatest technological achievements of the human race. The laws of probability make it clear that not even a single protein could ever have come into existence by mere chance; and if this is true of proteins–the most basic building-blocks of cells–the accidental formation of a complete cell is not even thinkable. This is of course proof of creation.

There are 2,000 types of proteins in a simple bacterium. The probability of their all coming into existence by accident is 1 in $10^{40.000}$. In a human being there are 200,000 types of proteins. The word "impossible" is too tame to describe the likelihood of such an event occurring by chance.

Because this is a subject that is discussed in more detail in other books of ours, we will just present a few simple examples here.

Earlier in this book we showed how the accidental formation of the balances that prevail in the universe was impossible. We will now show how the same is true for the accidental formation of even the simplest life-form. One study on this subject that we can refer to is a calculation made by Robert Shapiro, a professor of chemistry and expert on the subject of DNA at New York University. Shapiro, who is both a Darwinist and an evolutionist by the way, calculated the probability that all 2,000 of the different types of proteins that it takes to make up even a simple bacterium (the human body contains about 200,000 different types), could have come into being completely by chance. According to Shapiro, the probability is one in $10^{40.000}$.[102] (That number is "1" followed by forty thousand zeros. and it has no equivalent in the universe.)

Certainly it is plain what Shapiro's number must mean: The materialist (and its companion Darwinist) "explanation" that life evolved as an accident is certainly invalid. Chandra Wickramasinghe, a professor of applied mathematics and astronomy at the University of Cardiff commented on Shapiro's result:

> The likelihood of the spontaneous formation of life from inanimate matter is one to a number with $10^{40.000}$ noughts after it...It is big enough to bury Darwin and the whole theory of evolution. There was no primeval soup, neither on this planet nor on any other, and if the beginnings of life were not random, they must therefore have been the product of purposeful intelligence.[103]

The astronomer Fred Hoyle makes the same point:

> Indeed, such a theory (that life was assembled by an intelligence) is so obvious that one wonders why it is not widely accepted as being self- evident. The reasons are psychological rather than scientific.[104]

Both Wickramasinghe and Hoyle are men who, during much of their careers, approached science with a materialist bent; but the truth that confronted them was that life was created and both had the courage to admit this. Today, many more biologists and biochemists have put aside the fairy-tale that life could have emerged as an accident.

Those who are still loyal to Darwinism–those who still contend that life is a result of chance–are indeed in a state of consternation as we said at the beginning of this chapter. Just as the biochemist Michael Behe meant while he said, "The resulting realisation that life was designed by an intelligence is a shock to us in the twentieth century who have gotten used to thinking of life as the result of simple natural laws."[105], the shock that such people feel is the shock of having to come to terms with the reality of the existence of Allah, Who created them.

The dilemma that these adherents of materialism have fallen into was inevitable because they are struggling to deny a reality that they can clearly see. In the Qur'an, Allah describes the perplexity of those who believe in materialism like this:

By the Sky with its oscillating orbits. Most surely, you are at variance with each other in what you say. Averted from it is he who is averted. Cursed be the conjecturers; those who flounder in a glut of ignorance. (Surat adh-Dhariyat: 7-11)

At this point, our duty is to summon those who, influenced by materialist philosophy, have overstepped the bounds of reason, to reason and commonsense. We have to call them to cast aside all their prejudices and to think, to ponder the extraordinary design of the universe and of the life in it and to accept it as the plainest proof of the fact of Allah's creation.

But the real author of this call is not ourselves but Allah. Allah, Who created heaven and earth from nothing, summons the human beings He created to exercise their reason:

Your Lord is Allah, Who created the heavens and the earth in seven days and then established Himself firmly on the Throne. He directs the whole affair. No one can intercede except with His permission. That is Allah your Lord, so worship Him. Will you not pay heed? (SurahYunus: 3)

In another verse, mankind is addressed thus:

Is He who creates like him who does not create? So will you not pay heed? (Surat an-Nahl: 17)

Modern science has itself proven the truth of creation. It is now time for the scientific world to see this truth and derive a lesson from it. Those who deny or ignore the existence of Allah, and this is especially true of those who pretend that they are doing so in the name of science, should realize how deeply misled they are and turn away from this path.

On the other hand, this truth revealed by science has another lesson to teach to those who say that they already believed in the existence of Allah and that the universe was created by Him. The lesson is that their belief may be superficial and that they have not fully thought about the evidence of Allah's creation or about its consequences and that, for this reason, they may not be fulfilling all the responsibilities incumbent upon their belief. In the Qur'an, Allah describes such people like this:

Say: "To whom does the earth belong, and everyone in it, if you have any knowledge?"

They will say: "To Allah." Say: "So will you not pay heed?"

Say: "Who is the Lord of the Heavens and the Lord of the Mighty Throne? They will say: "Allah." Say: "So will you not have taqwa?"

Say: "In whose hand is the dominion over everything, He who gives protection and from whom no protection can be given, if you have any knowledge?"

They will say: "Allah's." Say: "So how have you been bewitched?" (Surat al-Muminun: 84-89)

Having come to the realization that Allah exists and that He created everything, to remain indifferent to this truth is indeed a sort of "bewitched". It is Allah Who created the universe and the world in which we live perfectly for us and then brought us into being as well. The duty of every person is to regard this as the most important fact of his life. Heaven and earth and everything in between belong to Allah the Sublime. Humanity should regard Allah as its Lord and Master and serve Him as is due. This is the truth revealed to us by Allah in the words:

He is the Lord of Heavens and the earth and everything in between them, so worship Him and persevere in His worship. Do you know of any other with His name? (Surah Maryam: 65)

The creation of the heavens and earth is far greater than the creation of mankind. But most of mankind do not know it.
(Surah Ghafir: 57)

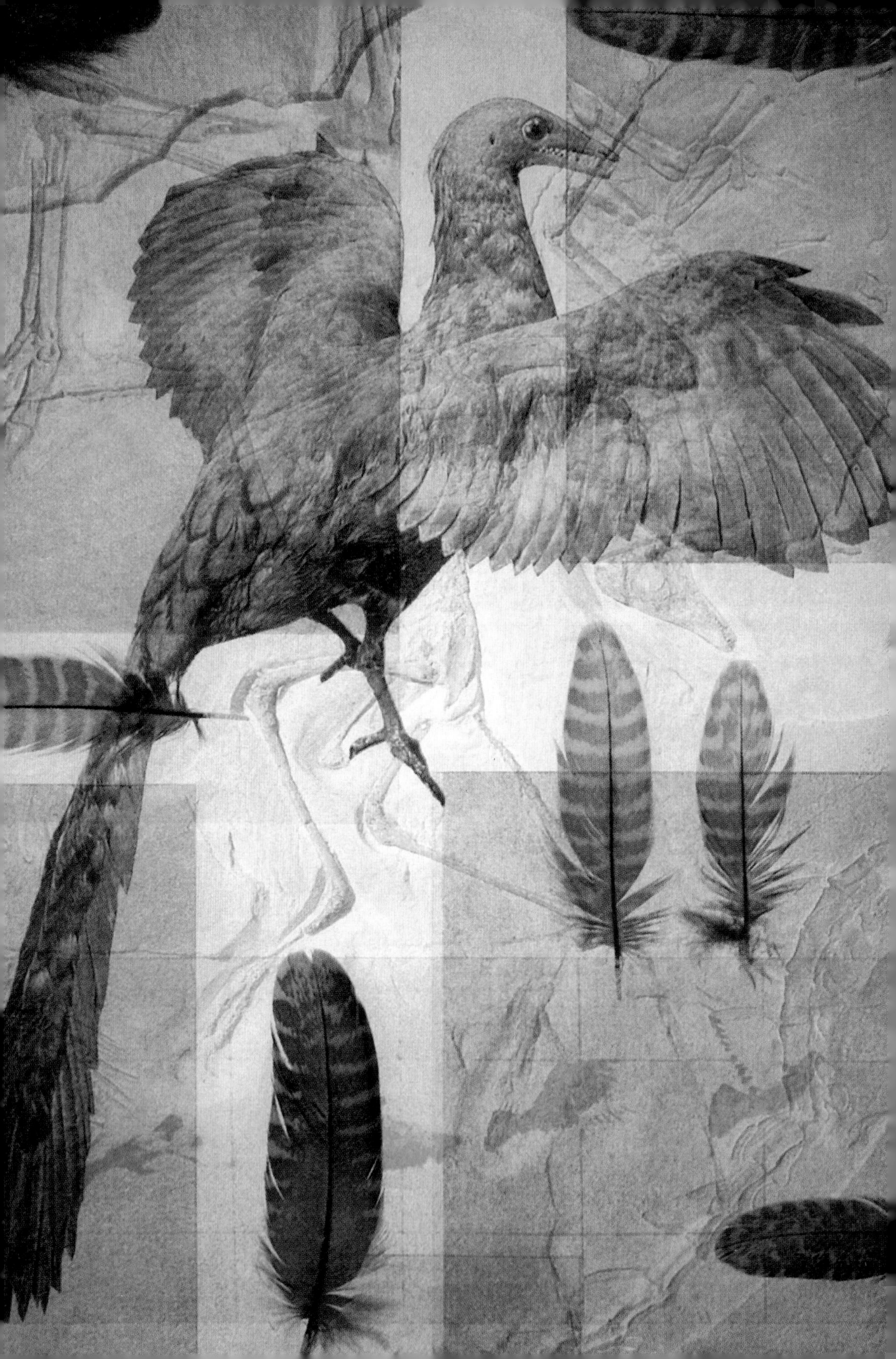

APPENDIX

THE EVOLUTION DECEIT

Question is: Can you tell me anything you know about evolution, any one thing that is true? I tried that question on the geology staff at the Field Museum of Natural History and the only answer I got was silence... Then I woke up and realized that all my life I had been duped into taking evolutionism as revealed truth in some way.

Colin Patterson, Senior Palaeontologist at the British Museum of Natural History and author of the book *Evolution*.[106]

Throughout this book, we focused our attention on non-living nature, namely on heavenly bodies, light, atoms and elements. After a thorough examination, we arrived to the conclusion that the universe can by no means be the product of coincidence. Rather, every and each detail of the universe shows a superior creation. Meanwhile, this conclusion verifies that materialism, in an endeavour to deny the creation in the universe, is nothing but a fallacy.

Invalidation of materialism surely renders all the other theories having their roots in this theory groundless. The foremost of these theories is Darwinism, or as differently called, the theory of evolution. This theory holding that life came into existence out of non-living beings actually collapsed with the fact that the universe is created by Allah. Hugh Ross, the American astrophysicist, explains this fact as follows:

> Atheism, Darwinism and virtually all the "isms" emanating from the eighteenth to twentieth century philosophies are built upon the assumption, the incorrect assumption, that the universe is infinite. The singularity (of the Big Bang) has brought us face to face with the Cause beyond / behind / before the universe and all that it contains, including life itself.[107]

Allah creates the universe and designs every detail of it. Therefore, it is impossible that the theory of evolution that attributes the very existence of living beings to coincidences can be true.

Indeed, when we analyse the evolution theory we see that scientific findings actually refute the theory. The design inherent in living beings is more brilliant and complex than the design of the non-living world we analysed throughout this book. In the world of living beings, we can analyse how atoms are arranged delicately. We can extend our analysis to see how the extraordinary mechanisms proteins, enzymes and cells possess.

This remarkable design in life has surely invalidated Darwinism at the end of the 20th century.

This issue is tackled in detail in our other works. However, due to the importance of the subject, we find it necessary to outline it below.

The Collapse of the Theory

The theory of evolution is a philosophy and a conception of the world that produces false hypotheses, assumptions and imaginary scenarios in order to explain the existence and origin of life in terms of mere coincidences. The roots of this philosophy go back as far as antiquity and ancient Greece.

All atheist philosophies that deny creation, directly or indirectly embrace and defend the idea of evolution. The same condition today applies to all the ideologies and systems that are antagonistic to religion.

The evolutionary notion has been cloaked in a scientific disguise for the last century and a half in order to justify itself. Though put forward as a supposedly scientific theory during the mid-19th century, the theory, despite all the best efforts of its advocates, has not so far been verified by any scientific finding or experiment. Indeed, the "very science" on which the theory depends so greatly has demonstrated and continues to demonstrate repeatedly that the theory has no merit in reality.

Laboratory experiments and probabilistic calculations have definitely made it clear that the amino acids from which life arises cannot have been formed by chance. The cell, which supposedly emerged by chance under primitive and uncontrolled terrestrial conditions according to evolutionists, still cannot be synthesised even in the most sophisticated, high-tech laboratories of the 20th century. Not a single "transitional form", creatures which are supposed to show the gradual evolution of advanced organisms from more primitive ones as neo-Darwinist theory claims, has ever been found anywhere in the world despite the most diligent and prolonged search in the fossil record.

Striving to gather evidence for evolution, evolutionists have unwittingly proven by their own hands that evolution cannot have happened at all!

The person who originally put forward the theory of evolution, essentially in the form that it is defended today, was an amateur English biologist by the name of Charles Robert Darwin. Darwin first published his ideas in a book entitled *The Origin of Species by Means of Natural Selection* in 1859. Darwin claimed in his book that all living beings had a common ancestor and that they evolved from one another by means of natural selection. Those that best adapted to the habitat transferred their traits to subsequent generations, and by accumulating over great epochs, these advantageous qualities transformed individuals into totally different species from their ancestors. The human being was thus the most developed product of the mechanism of natural selection. In short, the origin of one species was another species.

Darwin's fanciful ideas were seized upon and promoted by certain ideological and political circles and the theory became very popular. The main reason was that the level of knowledge of those days was not yet sufficient to reveal that Darwin's imaginary scenarios were false. When Darwin put forward his assumptions, the disciplines of genetics, microbiology, and biochemistry did not yet exist. If they had, Darwin might easily have recognised that his theory was totally unscientific and thus would not have attempted to advance such meaningless claims: the information determining species already exists in the genes and it is impossible for natural selection to produce new species by altering genes.

Charles Darwin

While the echoes of Darwin's book reverberated, an Austrian botanist by the name of Gregor Mendel discovered the laws of inheritance in 1865. Although little known before the end of the century, Mendel's discovery gained great importance in the early 1900s with the birth of the science of genetics. Some time later, the structures of genes and chromosomes were discovered. The discovery, in the 1950s, of the DNA molecule, which incorporates genetic information, threw the theory of evolution into a great crisis, because the origin of the immense amount of information in DNA could not possibly be explained by coincidental happenings.

Besides all these scientific developments, no transitional forms, which were supposed to show the gradual evolution of living organisms from primitive to advanced species, have ever been found despite years of search.

These developments ought to have resulted in Darwin's theory being banished to the dustbin of history. However, it was not, because certain circles insisted on revising, renewing, and elevating the theory to a scientific platform. These efforts gain meaning only if we realise that behind the theory lie ideological intentions rather than scientific concerns.

Nevertheless, some circles that believed in the necessity of upholding a theory that had reached an impasse soon set up a new model. The name of this new model was neo-Darwinism. According to this theory, species evolved as a result of mutations, minor changes in their genes, and the fittest ones survived through the mechanism of natural selection. When, however, it was proved that the mechanisms proposed by neo-Darwinism were invalid and minor changes were not sufficient for the formation of living beings, evolutionists went on to look for new models. They came up with a new claim called "punctuated equilibrium" that rests on no rational or scientific grounds. This model held that living beings suddenly evolved into another species without any transitional forms. In other words, species with no evolutionary "ancestors" suddenly appeared. This was a way of describing creation, though evolutionists would be loath to admit this. They tried to cover it up with incomprehensible scenarios. For instance, they said that the first bird in history could all of a sudden inexplicably have popped out of a reptile egg. The same theory also held that carnivorous land-dwelling animals could have turned into giant whales, having undergone a sudden and comprehensive transformation.

These claims, totally contradicting all the rules of genetics, biophysics, and biochemistry are as scientific as fairy-tales of frogs turning into princes! Nevertheless, being distressed by the crisis that the neo-Darwinist assertion was in, some evolutionist paleontologists embraced this theory, which has the distinction of being even more bizarre than neo-Darwinism itself.

The only purpose of this model was to provide an explanation for the gaps in the fossil record that the neo-Darwinist model could not explain. However, it is hardly rational to attempt to explain the gap in the fossil record of the evolution of birds with a claim that "a bird popped all of a sudden out of a reptile egg", because, by the evolutionists' own admission, the evolution of a species to another species requires a great and advantageous change in genetic information. However, no mutation whatsoever improves the genetic information or adds new information to it. Mutations only derange genetic information. Thus, the "gross mutations" imagined by the punctuated equilibrium model, would only cause "gross", that is "great", reductions and impairments in the genetic information.

The theory of punctuated equilibrium was obviously merely a product of the imagination. Despite this evident truth, the advocates of evolution did not hesitate to honour this theory. The fact that the model of evolution proposed by Darwin could not be proved by the fossil record forced them to do so. Darwin claimed that species underwent a gradual change, which necessitated the existence of half-bird/half-reptile or half-fish/half-reptile freaks. However, not even one of these "transitional forms" was found despite the extensive studies of evolutionists and the hundreds of thousands of fossils that were unearthed.

Evolutionists seized upon the model of punctuated equilibrium with the hope of concealing this great fossil fiasco. As we have stated before, it was very evident that this theory is a fantasy, so it very soon consumed itself. The model of punctuated equilibrium was never put forward as a consistent model, but rather used as an escape in cases that plainly did not fit the model of gradual evolution. Since evolutionists today realise that complex organs such as eyes, wings, lungs, brain and others explicitly refute the model of gradual evolution, in these particular points they are compelled to take shelter in the fantastic interpretations of the model of punctuated equilibrium.

Is There Any Fossil Record to Verify the Theory of Evolution?

The theory of evolution argues that the evolution of a species into another species takes place gradually, step-by-step over millions of years. The logical inference drawn from such a claim is that monstrous living organisms called "transitional forms" should have lived during these periods of transformation. Since evolutionists allege that all living things evolved from each other step-by-step, the number and variety of these transitional forms should have been in the millions.

If such creatures had really lived, then we should see their remains everywhere. In fact, if this thesis is correct, the number of intermediate transitional forms should be even greater than the number of animal species alive today and their fossilised remains should be abundant all over the world.

Since Darwin, evolutionists have been searching for fossils and the result has been for them a crushing disappointment. Nowhere in the world – neither on land nor in the depths of the sea – has any intermediate transitional form between any two species ever been uncovered.

Darwin himself was quite aware of the absence of such transitional forms. It was his greatest hope that they would be found in the future. Despite his hopefulness, he saw that the biggest stumbling block to his theory was the missing transitional forms. This is why, in his book *The Origin of Species*, he wrote:

> Why, if species have descended from other species by fine gradations, do we not everywhere see innumerable transitional forms? Why is not all nature in confusion, instead of the species being, as we see them, well defined?... But, as by this theory innumerable transitional forms must have existed, why do

> we not find them embedded in countless numbers in the crust of the earth?... But in the intermediate region, having intermediate conditions of life, why do we not now find closely-linking intermediate varieties? This difficulty for a long time quite confounded me.[108]

Darwin was right to be worried. The problem bothered other evolutionists as well. A famous British paleontologist, Derek V. Ager, admits this embarrassing fact:

> The point emerges that if we examine the fossil record in detail, whether at the level of orders or of species, we find – over and over again – not gradual evolution, but the sudden explosion of one group at the expense of another.[109]

The gaps in the fossil record cannot be explained away by the wishful thinking that not enough fossils have yet been unearthed and that these missing fossils will one day be found. Another evolutionist paleontologist, T. Neville George, explains the reason:

> There is no need to apologise any longer for the poverty of the fossil record. In some ways, it has become almost unmanageably rich and discovery is outpacing integration... The fossil record nevertheless continues to be composed mainly of gaps.[110]

Life Emerged on Earth Suddenly and in Complex Forms

When terrestrial strata and the fossil record are examined, it is seen that living organisms appeared simultaneously. The oldest stratum of the earth in which fossils of living creatures have been found is that of the "Cambrian", which has an estimated age of 530-520 million years.

Living creatures that are found in the strata belonging to the Cambrian period emerged in the fossil record all of a sudden without any pre-existing ancestors. The vast mosaic of living organisms, made up of such great numbers of complex creatures, emerged so suddenly that this miraculous event is referred to as the "Cambrian Explosion" in scientific literature.

Most of the organisms found in this stratum have highly advanced

'he Most Cherished Pieces of Evidence of Evolution are Proven to be Invalid

A four hundred and ten million-year-old Coelacanth fish fossil (below). Evolutionists claimed that it was the transitional form proving the transition of this fish from water to land. The fact that more than forty living examples of this fish have been caught in the last fifty years reveals that this is still a perfectly ordinary fish and that it is still living. A one hundred and thirty-five million-year-old Archaeopteryx fossil, the alleged ancestor of birds, which is said to have evolved from dinosaurs (right). Research on the fossil showed it, on the contrary, to be an extinct bird that had once flown but later lost that ability.

organs like eyes, or systems seen in organisms with a highly advanced organisation such as gills, circulatory systems, and so on. There is no sign in the fossil record to indicate that these organisms had any ancestors. Richard Monestarsky, the editor of *Earth Sciences* magazine, states about the sudden emergence of living species:

> A half-billion years ago the remarkably complex forms of animals that we see today suddenly appeared. This moment, right at the start of Earth's Cambrian Period, some 550 million years ago, marks the evolutionary explosion that filled the seas with the world's first complex creatures. The

> large animal phyla of today were present already in the early Cambrian and they were as distinct from each other then as they are today.[111]

Not being able to find answers to the question of how earth came to overflow with thousands of different animal species, evolutionists posit an imaginary period of 20 million years before the Cambrian Period to explain how life originated and "the unknown happened". This period is called the "evolutionary gap". No evidence for it has ever been found and the concept is still conveniently nebulous and undefined even today.

In 1984, numerous complex invertebrates were unearthed in Chengjiang, set in the central Yunnan plateau in the high country of southwest China. Among them were trilobites, now extinct, but no less complex in structure than any modern invertebrate.

The Swedish evolutionist paleontologist, Stefan Bengston, explains the situation as follows:

> If any event in life's history resembles man's creation myths, it is this sudden diversification of marine life when multicellular organisms took over as the dominant actors in ecology and evolution. Baffling (and embarrassing) to Darwin, this event still dazzles us.[112]

The sudden appearance of these complex living beings with no predecessors is no less baffling (and embarrassing) for evolutionists today than it was for Darwin 135 years ago. In nearly a century and a half, they have advanced not one step beyond the point that stymied Darwin.

As may be seen, the fossil record indicates that living things did not evolve from primitive to advanced forms, but instead emerged all of a sudden and in a perfect state. The absence of the transitional forms is not peculiar to the Cambrian period. Not a single transitional form verifying the alleged evolutionary "progression" of vertebrates – from fish to amphibians, reptiles, birds, and mammals – has ever been found. Every living species appears instantaneously and in its current form, perfect and complete, in the fossil record.

In other words, living beings did not come into existence through evolution. They were created.

EVOLUTION FORGERIES

Deceptions in Drawings

The fossil record is the principal source for those who seek evidence for the theory of evolution. When inspected carefully and without prejudice, the fossil record refutes the theory of evolution rather than supporting it. Nevertheless, misleading interpretations of fossils by evolutionists and their prejudiced representation to the public have given many people the impression that the fossil record indeed supports the theory of evolution.

The susceptibility of some findings in the fossil record to all kinds of interpretations is what best serves the evolutionists' purposes. The fossils unearthed are most of the time unsatisfactory for reliable identification. They usually consist of scattered, incomplete bone fragments. For this reason, it is very easy to distort the available data and to use it as desired. Not surprisingly, the reconstructions (drawings and models) made by evolutionists based on such fossil remains are prepared entirely speculatively in order to confirm evolutionary theses. Since people are readily affected by visual information, these imaginary reconstructed models are employed to convince them that the reconstructed creatures really existed in the past.

Evolutionist researchers draw human-like imaginary creatures, usually setting out from a single tooth, or a mandible fragment or a humerus, and present them to the public in a sensational manner as if they were links in human evolution. These drawings have played a great role in the establishment of the image of "primitive men" in the minds of many people.

These studies based on bone remains can only reveal very general characteristics of the creature concerned. The distinctive details are present in the soft tissues that quickly vanish with time. With the soft tissues speculatively interpreted, everything becomes possible within the boundaries of the imagination of the reconstruction's producer. Earnst A. Hooten from Harvard University explains the situation like this:

> To attempt to restore the soft parts is an even more hazardous undertaking.

> The lips, the eyes, the ears, and the nasal tip leave no clues on the underlying bony parts. You can with equal facility model on a Neanderthaloid skull the features of a chimpanzee or the lineaments of a philosopher. These alleged restorations of ancient types of man have very little if any scientific value and are likely only to mislead the public... So put not your trust in reconstructions.[113]

Studies Made to Fabricate False Fossils

Unable to find valid evidence in the fossil record for the theory of evolution, some evolutionists have ventured to manufacture their own. These efforts, which have even been included in encyclopaedias under the heading "evolution forgeries", are the most telling indication that the theory of evolution is an ideology and a philosophy that evolutionists are hard put to defend. Two of the most egregious and notorious of these forgeries are described below.

Piltdown Man

Charles Dawson, a well-known doctor and amateur paleoanthropologist, came forth with a claim that he had found a jawbone and a cranial fragment in a pit in the area of Piltdown, England, in 1912. Although the skull was human-like, the jawbone was distinctly simian. These specimens were christened the "Piltdown Man". Alleged to be 500 thousand years old, they were displayed as absolute proofs of human evolution. For more than 40 years, many scientific articles were written on the "Piltdown Man", many interpretations and drawings were made and the fossil was presented as crucial evidence of human evolution.

In 1949, scientists examined the fossil once more and concluded that the "fossil" was a deliberate forgery consisting of a human skull and the jawbone of an orang-utan.

Using the fluorine dating method, investigators discovered that the skull was only a few thousand years old. The teeth in the jawbone, which belonged to an orang-utan, had been artificially worn down and the "primitive" tools that had conveniently accompanied the fossils were crude forgeries that had been sharpened with steel implements. In the detailed analysis completed by Oakley, Weiner and Clark, they revealed this forgery to the public in 1953. The skull belonged to a 500-year-old man, and the mandibular bone belonged to a recently deceased ape! The teeth were thereafter specially arranged in an array and added to the jaw and the joints were filed in order to make them resemble that of a man. Then all these pieces were stained with potassium dichromate to give them a dated appearance. (These stains disappeared when dipped in acid.) Le Gros Clark, who was a member of the team that disclosed the forgery, could not hide his astonishment:

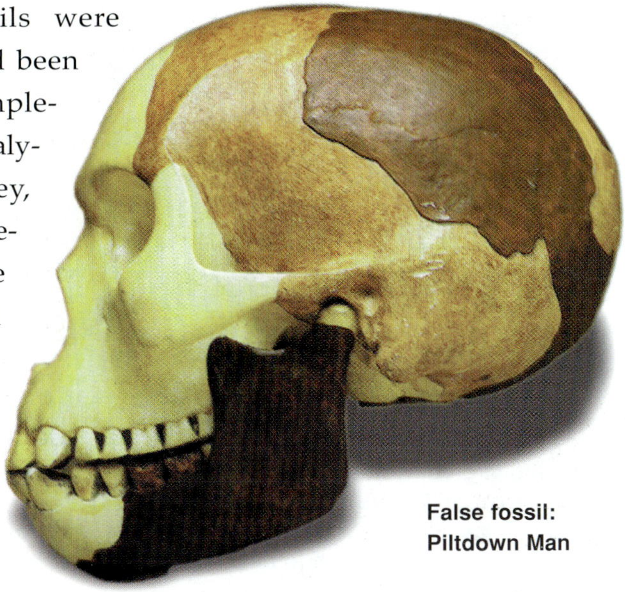

False fossil: Piltdown Man

> The evidences of artificial abrasion immediately sprang to the eye. Indeed so obvious did they seem it may well be asked: how was it that they had escaped notice before?[114]

Nebraska Man

In 1922, Henry Fairfield Osborn, the director of the American Museum of Natural History, declared that he had found a molar tooth fossil in western Nebraska near Snake Brook belonging to the Pliocene period. This tooth allegedly bore the common characteristics of both man and ape. Deep scientific arguments began in which some interpreted this tooth to be that of Pithecanthropus erectus while others claimed it was closer to that of modern human beings. This fossil, which aroused exten-

The above picture was drawn based on a single tooth and it was published in the Illustrated London News of 24th July 1922. However, evolutionists were extremely disappointed when it was revealed that this tooth belonged neither to an ape-like creature nor to a man, but to an extinct species of pig.

sive debate, was popularly named "Nebraska Man". It was also immediately given a "scientific name": "Hesperopithecus Haroldcooki".

Many authorities gave Osborn their support. Based on this single tooth, reconstructions of Nebraska Man's head and body were drawn. Moreover, Nebraska Man was even pictured with a whole family.

In 1927, other parts of the skeleton were also found. According to these newly discovered pieces, the tooth belonged neither to a man nor to an ape. It was realised that it belonged to an extinct species of wild American pig called Prosthennops.

Did Men and Apes Come from a Common Ancestor?

According to the claims of the theory of evolution, men and modern apes have common ancestors. These creatures evolved in time and some of them became the apes of today, while another group that followed another branch of evolution became the men of today.

Evolutionists call the so-called first common ancestors of men and apes "Australopithecus" which means "South African ape". Australopithecus, nothing but an old ape species that has become extinct, has various types. Some of them are robust, while others are small and slight.

Evolutionists classify the next stage of human evolution as "Homo", that is "man". According to the evolutionist claim, the living beings in the Homo series are more developed than Australopithecus, and not very much different from modern man. The modern man of our day, Homo sapiens, is said to have formed at the latest stage of the evolution of this species.

The fact of the matter is that the beings called Australopithecus in this imaginary scenario fabricated by evolutionists really are apes that became extinct, and the beings in the Homo series are members of various human races that lived in the past and then disappeared. Evolutionists arranged various ape and human fossils in an order from the smallest to the biggest in order to form a "human evolution" scheme. Research, however, has demonstrated that these fossils by no means imply an evolutionary process and some of these alleged ancestors of man were real apes whereas some of them were real humans.

Now, let us have a look at Australopithecus, which represents to evolutionists the first stage of the scheme of human evolution.

Australopithecus: Extinct Apes

Evolutionists claim that Australopithecus are the most primitive ancestors of modern men. These are an old species with a head and skull structure similar to that of modern apes, yet with a smaller cranial capacity. According to the claims of evolutionists, these creatures have a very important feature that authenticates them as the ancestors of men: bipedalism.

The movements of apes and men are completely different. Human beings are the only living creatures that move freely about on two feet. Some other animals do have a limited ability to move in this way, but those that do have bent skeletons.

According to evolutionists, these living beings called

Australopithecus had the ability to walk in a bent rather than an upright posture like human beings. Even this limited bipedal stride was sufficient to encourage evolutionists to project onto these creatures that they were the ancestors of man.

However, the first evidence refuting the allegations of evolutionists that Australopithecus were bipedal came from evolutionists themselves. Detailed studies made on Australopithecus fossils forced even evolutionists to admit that these looked "too" ape-like. Having conducted detailed anatomical research on Australopithecus fossils in the mid-1970s, Charles E. Oxnard likened the skeletal structure of Australopithecus to that of modern orang-utans:

> An important part of today's conventional wisdom about human evolution is based on studies of teeth, jaws and skull fragments of australopithecine fossils. These all indicate that the close relation of the australopithecine to the human lineage may not be true. All these fossils are different from gorillas, chimpanzees and men. Studied as a group, the australopithecine seems more like the orang-utan.[115]

What really embarrassed evolutionists was the discovery that Australopithecus could not have walked on two feet and with a bent posture. It would have been physically very ineffective for Australopithecus, allegedly bipedal but with a bent stride, to move about in such a way because of the enormous energy demands it would have entailed. By means of computer simulations conducted in 1996, the English paleoanthropologist Robin Crompton also demonstrated that such a "compound" stride was impossible. Crompton reached the following conclusion: a living being can walk either upright or on all fours. A type of in-between stride cannot be sustained for long periods because of the extreme energy consumption. This means that Australopithecus could not have been both bipedal and have a bent walking posture.

Probably the most important study demonstrating that Australopithecus could not have been bipedal came in 1994 from the research anatomist Fred Spoor and his team in the Department of Human Anatomy and Cellular Biology at the University of Liverpool, England.

This group conducted studies on the bipedalism of fossilised living beings. Their research investigated the involuntary balance mechanism found in the cochlea of the ear, and the findings showed conclusively that Australopithecus could not have been bipedal. This precluded any claims that Australopithecus was human-like.[93]

The Homo Series: Real Human Beings

The next step in the imaginary human evolution is "Homo", that is, the human series. These living beings are humans who are no different from modern men, yet who have some racial differences. Seeking to exaggerate these differences, evolutionists represent these people not as a "race" of modern man but as a different "species". However, as we will soon see, the people in the Homo series are nothing but ordinary human racial types.

According to the fanciful scheme of evolutionists, the internal imaginary evolution of the Homo species is as follows: First Homo erectus, then Homo sapiens archaic and Neanderthal Man, later Cro-Magnon Man and finally modern man.

Despite the claims of evolutionists to the contrary, all the "species" we have enumerated above are nothing but genuine human beings. Let us first examine Homo erectus, who evolutionists refer to as the most primitive human species.

The most striking evidence showing that Homo erectus is not a "primitive" species is the fossil of "Turkana Boy", one of the oldest Homo erectus remains. It is estimated that the fossil was of a 12-year-old boy, who would have been 1.83 meters tall in his adolescence. The upright skeletal structure of the fossil is no different from that of modern man. Its tall and slender skeletal structure totally complies with that of the people living in tropical regions in our day. This fossil is one of the most important pieces of evidence that Homo erectus is simply another specimen of the modern human race. Evolutionist paleontologist Richard Leakey compares Homo erectus and modern man as follows:

> One would also see differences in the shape of the skull, in the degree of protrusion of the face, the robustness of the brows and so on. These differences are probably no more pronounced than we see today between the separate geographical races of modern humans. Such biological variation arises when populations are geographically separated from each other for significant lengths of time.[116]

Leakey means to say that the difference between Homo erectus and us is no more than the difference between Negroes and Eskimos. The cranial features of Homo erectus resulted from their manner of feeding, and genetic emigration and from their not assimilating with other human races for a lengthy period.

Another strong piece of evidence that Homo erectus is not a "primitive" species is that fossils of this species have been unearthed aged twenty-seven thousand years and even thirteen thousand years. According to an article published in Time – which is not a scientific periodical, but nevertheless had a sweeping effect on the world of science – Homo erectus fossils aged twenty-seven thousand years were found on the island of Java. In the Kow swamp in Australia, some thirteen thousand year-old fossils were found that bore Homo Sapiens-Homo Erectus characteristics. All these fossils demonstrate that Homo erectus continued living up to times very close to our day and were nothing but a human race that has since been buried in history.

Archaic Homo Sapiens and Neanderthal Man

Archaic Homo sapiens is the immediate forerunner of contemporary man in the imaginary evolutionary scheme. In fact, evolutionists do not have much to say about these men, as there are only minor differences between them and modern men. Some researchers even state that representatives of this race are still living today, and point to the Aborigines in Australia as an example. Like Homo sapiens, the Aborigines also have thick protruding eyebrows, an inward-inclined mandibular structure, and a slightly smaller cranial volume. Moreover, significant discoveries have been made hinting that such people lived in Hungary and in some villages in Italy until not very long ago.

Evolutionists point to human fossils unearthed in the Neander valley of Holland which have been named Neanderthal Man. Many contemporary researchers define Neanderthal Man as a sub-species of modern man and call it "Homo sapiens neandertalensis". It is definite that this race lived together with modern humans, at the same time and in the same areas. The findings testify that Neanderthals buried their dead, fashioned musical instruments, and had cultural affinities with the Homo sapiens sapiens living during the same period. Entirely modern skulls and skeletal structures of Neanderthal fossils are not open to any speculation. A prominent authority on the subject, Erik Trinkaus from New Mexico University writes:

> Detailed comparisons of Neanderthal skeletal remains with those of modern humans have shown that there is nothing in Neanderthal anatomy that conclusively indicates locomotor, manipulative, intellectual, or linguistic abilities inferior to those of modern humans.[117]

In fact, Neanderthals even had some "evolutionary" advantages over modern men. The cranial capacity of Neanderthals was larger than that of the modern man and they were more robust and muscular than we are. Trinkaus adds: "One of the most characteristic features of the Neanderthals is the exaggerated massiveness of their trunk and limb bones. All of the preserved bones suggest a strength seldom attained by modern humans. Furthermore, not only is this robustness present among the adult males, as one might expect, but it is also evident in the adult females, adolescents, and even children."

To put it precisely, Neanderthals are a particular human race that assimilated with other races in time.

All of these factors show that the scenario of "human evolution" fabricated by evolutionists is a figment of their imaginations, and that men have always been men and apes always apes.

Can Life Result from Coincidences as Evolution Argues?

The theory of evolution holds that life started with a cell that formed by chance under primitive earth conditions. Let us therefore examine the composition of the cell with simple comparisons in order to show how irrational it is to ascribe the existence of the cell – a structure which still maintains its mystery in many respects, even at a time when we are about to set foot in the 21st century – to natural phenomena and coincidences.

With all its operational systems, systems of communication, transportation and management, a cell is no less complex than any city. It contains power stations producing the energy consumed by the cell, factories manufacturing the enzymes and hormones essential for life, a databank where all necessary information about all products to be produced is recorded, complex transportation systems and pipelines for carrying raw materials and products from one place to another, advanced laboratories and refineries for breaking down imported raw materials into their usable parts, and specialised cell membrane proteins for the control of incoming and outgoing materials. These constitute only a small part of this incredibly complex system.

Far from being formed under primitive earth conditions, the cell, which in its composition and mechanisms is so complex, cannot be synthesised in even the most sophisticated laboratories of our day. Even with the use of amino acids, the building blocks of the cell, it is not possible to produce so much as a single organelle of the cell, such as mitochondria or ribosome, much less a whole cell. The first cell claimed to have been produced by evolutionary coincidence is as much a figment of the imagination and a product of fantasy as the unicorn.

Proteins Challenge Coincidence

And it is not just the cell that cannot be produced: the formation, under natural conditions, of even a single protein of the thousands of complex protein molecules making up a cell is impossible.

Proteins are giant molecules consisting of amino acids arranged in a particular sequence in certain quantities and structures. These molecules constitute the building blocks of a living cell. The simplest is composed of 50 amino acids; but there are some proteins that are composed of thousands of amino acids. The absence, addition, or replacement of a single amino acid in the structure of a protein in living cells, each of which has a particular function, causes the protein to become a useless molecular heap. Incapable of demonstrating the "accidental formation" of amino acids, the theory of evolution founders on the point of the formation of proteins.

We can easily demonstrate, with simple probability calculations anybody can understand, that the functional structure of proteins can by no means come about by chance.

There are twenty different amino acids. If we consider that an average-sized protein molecule is composed of 288 amino acids, there are 10^{300} different combinations of acids. Of all of these possible sequences, only "one" forms the desired protein molecule. The other amino-acid chains are either completely useless or else potentially harmful to living things. In other words, the probability of the coincidental formation of only one protein molecule cited above is "1 in 10^{300}". The probability of this "1" occurring out of an "astronomical" number consisting of 1 followed by 300 zeros is for all practical purposes zero; it is impossible. Furthermore, a protein molecule of 288 amino acids is rather a modest one compared with some giant protein molecules consisting of thousands of amino acids. When we apply similar probability calculations to these giant protein molecules, we see that even the word "impossible" becomes inadequate.

If the coincidental formation of even one of these proteins is impossible, it is billions of times more impossible for approximately one million of those proteins to come together by chance in an organised fashion and make up a complete human cell. Moreover, a cell is not merely a collection of proteins. In addition to proteins, cells also include nucleic acids, carbohydrates, lipids, vitamins, and many other chemicals such as electrolytes, all of which are arranged harmoniously and with design in specific proportions, both in terms of structure and function. Each functions as a building block or component in various organelles.

As we have seen, evolution is unable to explain the formation of even a single protein out of the millions in the cell, let alone explain the cell.

Prof. Dr. Ali Demirsoy, one of the foremost authorities of evolutionist thought in Turkey, in his book *Kalitim ve Evrim* (Inheritance and Evolution), discusses the probability of the accidental formation of Cytochrome-C, one of the essential enzymes for life:

> The probability of the formation of a Cytochrome-C sequence is as likely as zero. That is, if life requires a certain sequence, it can be said that this has a probability likely to be realised once in the whole universe. Otherwise, some metaphysical powers beyond our definition should have acted in its formation. To accept the latter is not appropriate to the goals of science. We therefore have to look into the first hypothesis.[118]

After these lines, Demirsoy admits that this probability, which he accepted just because it was "more appropriate to the goals of science", is unrealistic:

> The probability of providing the particular amino acid sequence of Cytochrome-C is as unlikely as the possibility of a monkey writing the history of humanity on a typewriter – taking it for granted that the monkey pushes the keys at random.[119]

The correct sequence of proper amino acids is simply not enough for the formation of one of the protein molecules present in living things. Besides this, each of the twenty different types of amino acid present in the composition of proteins must be left-handed. Chemically, there are two different types of amino acids called "left-handed" and "right-handed". The difference between them is the mirror-symmetry between their three dimensional structures, which is similar to that of a person's right and left hands. Amino acids of either of these two types are found in equal numbers in nature and they can bond perfectly well with one another. Yet, research uncovers an astonishing fact: all proteins present in the structure of living things are made up of left-handed amino acids. Even a single right-handed amino acid attached to the structure of a protein renders it useless.

Let us for an instant suppose that life came into existence by chance as evolutionists claim. In this case, the right and left-handed amino acids that were generated by chance should be present in nature in roughly equal amounts. The question of how proteins can pick out only left-handed amino acids, and how not even a single right-handed amino acid becomes involved in the life process is something that still confounds evolutionists. In the *Britannica Science Encyclopaedia*, an ardent defender of evolution, the authors indicate that the amino acids of all living organisms on earth and the building blocks of complex polymers such as proteins have the same left-handed asymmetry. They add that this is tantamount to tossing a coin a million times and always getting heads. In the same encyclopaedia, they state that it is not possible to understand why molecules become left-handed or right-handed and that this choice is fascinatingly related to the source of life on earth.[120]

It is not enough for amino acids to be arranged in the correct numbers, sequences, and in the required three-dimensional structures. The formation of a protein also requires that amino acid molecules with more than one arm be linked to each other only through certain arms. Such a bond is called a "peptide bond". Amino acids can make different bonds with each other; but proteins comprise those and only those amino acids that join together by "peptide" bonds.

Research has shown that only 50 % of amino acids, combining at random, combine with a peptide bond and that the rest combine with different bonds that are not present in proteins. To function properly, each amino acid making up a protein must join with other amino acids with a peptide bond, as it has only to be chosen from among the left-handed ones. Unquestionably, there is no control mechanism to select and leave out the right-handed amino acids and personally make sure that each amino acid makes a peptide bond with the other.

Under these circumstances, the probabilities of an average protein molecule comprising five hundred amino acids arranging itself in the correct quantities and in sequence, in addition to the probabilities of all of the amino acids it contains being only left-handed and combining using only peptide bonds are as follows:

– The probability of being in the right sequence

= $1/20^{500}$ = $1/10^{650}$

– The probability of being left-handed

= $1/2^{500}$ = $1/10^{150}$

– The probability of combining using a "peptide bond"

= $1/2^{499}$ = $1/10^{150}$

TOTAL PROBABILITY

= $1/10^{950}$ that is, "1" probability in 10^{950}

As you can see above, the probability of the formation of a protein molecule comprising five hundred amino acids is "1" divided by a number formed by placing 950 zeros after a 1, a number incomprehensible to the human mind. This is only a probability on paper. Practically, such a possibility has "0" chance of realisation. In mathematics, a probability

The probability of an average protein molecule comprising five hundred amino acids being arranged in the correct proportion and sequence in addition to the probability of all of the amino acids it contains being only left-handed and being combined only with peptide bonds is "1" divided by 10^{950}. We can write this number, which is formed by putting 950 zeros after 1, as follows:

10^{950} =

**100,000,000,000,000,000,000,000,000,000,000,000,000,000,000,000,000,000,000,
000,000,000,000,000,000,000,000,000,000,000,000,000,000,000,000,000,000,000,
000,000,000,000,000,000,000,000,000,000,000,000,000,000,000,000,000,000,000,
000,000,000,000,000,000,000,000,000,000,000,000,000,000,000,000,000,000,000,
000,000,000,000,000,000,000,000,000,000,000,000,000,000,000,000,000,000,000,
000,000,000,000,000,000,000,000,000,000,000,000,000,000,000,000,000,000,000,
000,000,000,000,000,000,000,000,000,000,000,000,000,000,000,000,000,000,000,
000,000,000,000,000,000,000,000,000,000,000,000,000,000,000,000,000,000,000,
000,000,000,000,000,000,000,000,000,000,000,000,000,000,000,000,000,000,000,
000,000,000,000,000,000,000,000,000,000,000,000,000,000,000,000,000,000,000,
000,000,000,000,000,000,000,000,000,000,000,000,000,000,000,000,000,000,000,
000,000,000,000,000,000,000,000,000,000,000,000,000,000,000,000,000,000,000,
000,000,000,000,000,000,000,000,000,000,000,000,000,000,000,000,000,000,000,
000,000,000,000,000,000,000,000,000,000,000,000,000,000,000,000,000,000,000,
000,000,000,000,000,000,000,000,000,000,000,000,000,000,000,000,000,000,000,
000,000,000,000,000,000,000,000,000,000,000,000,000,000,000,000,000,000,000,
000,000,000,000,000,000,000,000,000,000,000,000,000**

smaller than 1 over 10^{50} is statistically considered to have a "0" probability of realisation.

While the improbability of the formation of a protein molecule made up of five hundred amino acids reaches such an extent, we can further proceed to push the limits of the mind to higher levels of improbability. In the "haemoglobin" molecule, a vital protein, there are five hundred and seventy-four amino acids, which is a much larger number than that of the amino acids making up the protein mentioned above. Now consider this: in only one out of the billions of red blood cells in your body, there are "280,000,000" (280 million) haemoglobin molecules. The supposed age of the earth is not sufficient to afford the formation of even a single protein, let alone a red blood cell, by the method of "trial and error". The conclusion from all this is that evolution falls into a terrible abyss of improbability right at the stage of the formation of a single protein.

Looking for Answers to the Generation of Life

Well aware of the terrible odds against the possibility of life forming by chance, evolutionists were unable to provide a rational explanation for their beliefs, so they set about looking for ways to demonstrate that the odds were not so unfavourable.

They designed a number of laboratory experiments to address the question of how life could generate itself from non-living matter. The best known and most respected of these experiments is the one known as the "Miller Experiment" or "Urey-Miller Experiment", which was conducted by the American researcher Stanley Miller in 1953.

With the purpose of proving that amino acids could have come into existence by accident, Miller created an atmosphere in his laboratory that he assumed would have existed on primordial earth (but which later proved to be unrealistic) and he set to work. The mixture he used for this primordial atmosphere was composed of ammonia, methane, hydrogen, and water vapour.

Miller knew that methane, ammonia, water vapour and hydrogen would not react with each other under natural conditions. He was aware that he had to inject energy into the mixture to start a reaction. He suggested that this energy could have come from lightning flashes in the primordial atmosphere and, relying on this supposition, he used an artificial electricity discharge in his experiments.

Miller boiled this gas mixture at 100°C for a week, and, in addition, he introduced an electric current into the chamber. At the end of the week, Miller analysed the chemicals that had been formed in the chamber and observed that three of the twenty amino acids, which constitute the basic elements of proteins, had been synthesised.

This experiment aroused great excitement among evolutionists and they promoted it as an outstanding success. Encouraged by the thought that this experiment definitely verified their theory, evolutionists immediately produced new scenarios. Miller had supposedly proved that amino acids could form by themselves. Relying on this, they hurriedly hypothesised the following stages. According to their scenario, amino acids had later by accident united in the proper sequences to form proteins. Some of these accidentally formed proteins placed themselves in cell membrane-like structures, which "somehow" came into existence and formed a primitive cell. The cells united in time and formed living organisms. The greatest mainstay of the scenario was Miller's experiment.

However, Miller's experiment was nothing but make-believe, and has since been proven invalid in many respects.

The Invalidity of Miller's Experiment

Nearly half a century has passed since Miller conducted his experiment. Although it has been shown to be invalid in many respects, evolutionists still advance Miller and his results as absolute proof that life could have formed spontaneously from non-living matter. When we assess Miller's experiment critically, without the bias and subjectivity of evolutionist thinking, however, it is evident that the situation is not as rosy as evolutionists would have us think. Miller set for himself the goal of proving that amino acids could form by themselves in earth's primitive conditions. Some amino acids were produced, but the conduct of the experiment conflicts with his goal in many ways, as we shall now see.

◆ Miller isolated the amino acids from the environment as soon as they were formed, by using a mechanism called a "cold trap". Had he not done so, the conditions of the environment in which the amino acids formed would immediately have destroyed the molecules.

It is quite meaningless to suppose that some conscious mechanism of this sort was integral to earth's primordial conditions, which involved ultraviolet radiation, thunderbolts, various chemicals, and a high percentage of free oxygen. Without such a mechanism, any amino acid that did manage to form would immediately have been destroyed.

◆ The primordial atmospheric environment that Miller attempted to simulate in his experiment was not realistic. Nitrogen and carbon dioxide would have been constituents of the primordial atmosphere, but Miller disregarded this and used methane and ammonia instead.

Why? Why were evolutionists insistent on the point that the primitive atmosphere contained high amounts of methane (CH_4), ammonia (NH_3), and water vapour (H_2O)? The answer is simple: without ammonia, it is impossible to synthesise an amino acid. Kevin McKean talks about this in an article published in Discover magazine:

> Miller and Urey imitated the ancient atmosphere of earth with a mixture of methane and ammonia. According to them, the earth was a true homogeneous mixture of metal, rock and ice. However in the latest studies, it is understood that the earth was very hot at those times and that it was composed of melted nickel and iron. Therefore, the chemical atmosphere of that time should have been formed mostly of nitrogen (N_2), carbon dioxide (CO_2) and water vapour (H_2O). However these are not as appropriate as methane and ammonia for the production of organic molecules.[121]

After a long period of silence, Miller himself also confessed that the atmospheric environment he used in his experiment was not realistic.

◆ Another important point invalidating Miller's experiment is that there was enough oxygen to destroy all the amino acids in the atmosphere at the time when evolutionists thought that amino acids formed. This oxygen concentration would definitely have hindered the formation of amino acids. This situation completely negates Miller's experiment, in which he totally neglected oxygen. If he had used oxygen in the experiment, methane would have decomposed into carbon dioxide and water, and ammonia would have decomposed into nitrogen and water.

On the other hand, since no ozone layer yet existed, no organic molecule could possibly have lived on earth because it was entirely unprotected against intense ultraviolet rays.

◆ In addition to a few amino acids essential for life, Miller's experiment also produced many organic acids with characteristics that are quite detrimental to the structures and functions of living things. If he had not isolated the amino acids and had left them in the same environment with these chemicals, their destruction or transformation into different compounds through chemical reactions would have been unavoidable. Moreover, a large number of right-handed amino acids also formed. The existence of these amino acids alone refuted the theory, even within its own reasoning, because right-handed amino acids are unable to function in the composition of living organisms and render proteins useless when they are involved in their composition.

To conclude, the circumstances in which amino acids formed in Miller's experiment were not suitable for life forms to come into being. The medium in which they formed was an acidic mixture that destroyed and oxidised any useful molecules that might have been obtained.

Evolutionists themselves actually refute the theory of evolution, as they are often wont to do, by advancing this experiment as "proof". If the experiment proves anything, it is that amino acids can only be produced in a controlled laboratory environment where all the necessary conditions have been specifically and consciously designed. That is, the experiment shows that what brings life (even the "near-life" of amino acids) into being cannot be unconscious chance, but rather conscious will – in a word, Creation. This is why every stage of Creation is a sign proving to us the existence and might of Allah.

The Miraculous Molecule: DNA

The theory of evolution has been unable to provide a coherent explanation for the existence of the molecules that are the basis of the cell. Furthermore, developments in the science of genetics and the discovery of the nucleic acids (DNA and RNA) have produced brand-new problems for the theory of evolution.

In 1955, the work of two scientists on DNA, James Watson and Francis Crick, launched a new era in biology. Many scientists directed their attention to the science of genetics. Today, after years of research, scientists have, largely, mapped the structure of DNA.

Here, we need to give some very basic information on the structure and function of DNA:

The molecule called DNA, which exists in the nucleus of each of the 100 trillion cells in our body, contains the complete construction plan of the human body. Information regarding all the characteristics of a person, from the physical appearance to the structure of the inner organs, is recorded in DNA by means of a special coding system. The information in DNA is coded within the sequence of four special bases that make up this molecule. These bases are specified as A, T, G, and C according to the initial letters of their names. All the structural differences among people depend on the variations in the sequence of these bases. There are approximately 3.5 billion nucleotides, that is, 3.5 billion letters in a DNA molecule.

The DNA data pertaining to a particular organ or protein is included in special components called "genes". For instance, information about the eye exists in a series of special genes, whereas information about the heart exists in quite another series of genes. The cell produces proteins by using the information in all of these genes. Amino acids that constitute the structure of the protein are defined by the sequential arrangement of three nucleotides in the DNA.

At this point, an important detail deserves attention. An error in the sequence of nucleotides making up a gene renders the gene completely

The molecule called DNA contains the complete construction plan of the human body.

useless. When we consider that there are 200 thousand genes in the human body, it becomes more evident how impossible it is for the millions of nucleotides making up these genes to form by accident in the right sequence. An evolutionist biologist, Frank Salisbury, comments on this impossibility by saying:

> A medium protein might include about 300 amino acids. The DNA gene controlling this would have about 1,000 nucleotides in its chain. Since there are four kinds of nucleotides in a DNA chain, one consisting of 1,000 links could exist in 4^{1000} forms. Using a little algebra (logarithms), we can see that $4^{1000}=10^{600}$. Ten multiplied by itself 600 times gives the figure 1 followed by 600 zeros! This number is completely beyond our comprehension.[122]

The number 4^{1000} is equivalent to 10^{600}. We obtain this number by adding 600 zeros to 1. As 10 with 11 zeros indicates a trillion, a figure with 600 zeros is indeed a number that is difficult to grasp.

Evolutionist Prof. Ali Demirsoy was forced to make the following admission on this issue:

> In fact, the probability of the random formation of a protein and a nucleic acid (DNA-RNA) is inconceivably small. The chances against the emergence of even a particular protein chain are astronomic.[123]

In addition to all these improbabilities, DNA can barely be involved in a reaction because of its double-chained spiral shape. This also makes it impossible to think that it can be the basis of life.

Moreover, while DNA can replicate only with the help of some enzymes that are actually proteins, the synthesis of these enzymes can be realised only by the information coded in DNA. As they both depend on each other, either they have to exist at the same time for replication, or one of them has had to be "created" before the other. American microbiologist Jacobson comments on the subject:

The complete directions for the reproduction of plans, for energy and the extraction of parts from the current environment, for the growth sequence, and for the effector mechanism translating instructions into growth – all had to be simultaneously present at that moment (when life began). This combination of events has seemed an incredibly unlikely happenstance, and has often been ascribed to divine intervention.[124]

The quotation above was written two years after the disclosure of the structure of DNA by James Watson and Francis Crick. Despite all the developments in science, this problem remains unsolved for evolutionists. To sum up, the need for DNA in reproduction, the necessity of the presence of some proteins for reproduction, and the requirement to produce these proteins according to the information in the DNA entirely demolish evolutionist theses.

Two German scientists, Junker and Scherer, explained that the synthesis of each of the molecules required for chemical evolution, necessitates distinct conditions, and that the probability of the compounding of these materials having theoretically very different acquirement methods is zero:

> Until now, no experiment is known in which we can obtain all the molecules necessary for chemical evolution. Therefore, it is essential to produce various molecules in different places under very suitable conditions and then to carry them to another place for reaction by protecting them from harmful elements like hydrolysis and photolysis.[125]

In short, the theory of evolution is unable to prove any of the evolutionary stages that allegedly occur at the molecular level.

To summarise what we have said so far, neither amino acids nor their products, the proteins making up the cells of living beings, could ever be produced in any so-called "primitive atmosphere" environment. Moreover, factors such as the incredibly complex structure of proteins, their right-hand, left-hand features, and the difficulties in the formation of peptide bonds are just parts of the reason why they will never be produced in any future experiment either.

Even if we suppose for a moment that proteins somehow did form accidentally, that would still have no meaning, for proteins are nothing at all on their own: they cannot themselves reproduce. Protein synthesis is only possible with the information coded in DNA and RNA molecules. Without DNA and RNA, it is impossible for a protein to reproduce. The specific sequence of the twenty different amino acids encoded in DNA determines the structure of each protein in the body. However, as has been made abundantly clear by all those who have studied these molecules, it is impossible for DNA and RNA to form by chance.

The Fact of Creation

With the collapse of the theory of evolution in every field, prominent names in the discipline of microbiology today admit the fact of creation and have begun to defend the view that everything is created by a conscious Creator as part of an exalted creation. This is already a fact that people cannot disregard. Scientists who can approach their work with an open mind have developed a view called "intelligent design". Michael J. Behe, one of the foremost of these scientists, states that he accepts the absolute being of the Creator and describes the impasse of those who deny this fact:

> The result of these cumulative efforts to investigate the cell – to investigate life at the molecular level – is a loud, clear, piercing cry of "design!" The result is so unambiguous and so significant that it must be ranked as one of the greatest achievements in the history of science. This triumph of science should evoke cries of "Eureka" from ten thousand throats.

> But, no bottles have been uncorked, no hands clapped. Instead, a curious, embarrassed silence surrounds the stark complexity of the cell. When the subject comes up in public, feet start to shuffle, and breathing gets a bit laboured. In private people are a bit more relaxed; many explicitly admit the obvious but then stare at the ground, shake their heads, and let it go like that. Why does the scientific community not greedily embrace its startling discovery? Why is the observation of design handled with intellectual gloves? The dilemma is that while one side of the elephant is labelled intelligent design, the other side must be labelled God.[126]

Today, many people are not even aware that they are in a position of accepting a body of fallacy as truth in the name of science, instead of believing in Allah. Those who do not find the sentence "Allah created you from nothing" scientific enough can believe that the first living being came into being by thunderbolts striking a "primordial soup" billions of years ago.

As we have described elsewhere in this book, the balances in nature are so delicate and so numerous that it is entirely irrational to claim that they developed "by chance". No matter how much those who cannot set themselves free from this irrationality may strive, the signs of Allah in the heavens and the earth are completely obvious and they are undeniable.

Allah is the Creator of the heavens, the earth and all that is in between.

The signs of His being have encompassed the entire universe.

Glory to You, of knowledge we have none, save what You have taught us: In truth it is You Who are perfect in knowledge and wisdom.

(Surat al-Baqara, 32)

Among His Signs is the creation of the heavens and earth and all the creatures He has spread about in them. And He has the power to gather them together whenever He wills.
(Surat ash-Shura: 29)

NOTES:

1 Arthur Koestler, *Janus: A Summing Up*, New York: Vintage Books, 1978, p. 250.
2 Andrei Linde, "The Self-Reproducing Inflationary Universe", *Scientific American*, vol. 271, 1994, p. 48
3 George Politzer, *Principes Fondamentaux de Philosophie*, Editions Sociales, Paris 1954 ,p. 84
4 S. Jaki, *Cosmos and Creator*, Regnery Gateway, Chicago, 1980, p. 54
5 Stephen Hawking, *Evreni Kucaklayan Karinca*, Alkim Publishing, 1993, p. 62-63
6 Henry Margenau, Roy Abraham Vargesse. *Cosmos, Bios, Theos*. La Salle IL: Open Court Publishing, 1992, p. 241
7 Hugh Ross, *The Creator and the Cosmos: How Greatest Scientific Discoveries of The Century Reveal God*, Colorado: NavPress, revised edition, 1995, p. 76
8 William Lane Craig, Cosmos and Creator, *Origins & Design*, Spring 1996, vol. 17, p. 19
9 William Lane Craig, Cosmos and Creator, *Origins & Design*, Spring 1996, vol. 17, p. 19
10 William Lane Craig, Cosmos and Creator, *Origins & Design*, Spring 1996, vol. 17, p. 20
11 Christopher Isham, "Space, Time and Quantum Cosmology", paper presented at the conference "God, Time and Modern Physics", March 1990, *Origins & Design*, Spring 1996, vol. 17, p. 27
12 R. Brout, Ph. Spindel, "Black Holes Dispute", *Nature*, vol 337, 1989, p. 216
13 Herbert Dingle, *Science at the Crossroads*, London: Martin Brian & O'Keefe, 1972, p. 31-32
14 StephenHawking, *A Brief History of Time*, New York: Bantam Books, 1988, p. 46
15 John Maddox, "Down with the Big Bang", *Nature*, vol. 340, 1989, p. 378
16 H. P. Lipson, "A Physicist Looks at Evolution", *Physics Bulletin*, vol. 138, 1980, p. 138
17 Paul Davies, *Superforce: The Search for a Grand Unified Theory of Nature*, 1984, p. 184
18 Fred Hoyle, *The Intelligent Universe*, London, 1984, p. 184-185
19 Paul Davies, *Superforce: The Search for a Grand Unified Theory of Nature*, 1984, p. 184
20 *Bilim ve Teknik* (Science and Technics) 201, p. 16
21 Stephen Hawking, *A Brief History Of Time*, Bantam Press, London: 1988, p. 121-125
22 Paul Davies. *God and the New Physics*. New York: Simon & Schuster, 1983, p. 189
23 Michael Denton, *Nature's Destiny: How the Laws of Biology Reveal Purpose in the Universe*, The New York: The Free Press, 1998, p. 12-13
24 Paul Davies. *The Accidental Universe*, Cambridge: Cambridge University Press, 1982, Foreword.
25 Hugh Ross, *The Creator and the Cosmos*, p. 122-23
26 Roger Penrose, *The Emperor's New Mind*, 1989; Michael Denton, *Nature's Destiny*, The New York: The Free Press, 1998, p. 9
27 George Greenstein, *The Symbiotic Universe*, p. 27
28 Hugh Ross, *Design and the Anthropic Principle, Reasons To Believe*, CA, 1988
29 Hugh Ross, *The Creator and the Cosmos*, p. 123
30 Paul Davies, *The Cosmic Blueprint*, London: Penguin Books, 1987, p. 203
31 Paul Davies, *Superforce*, New York: Simon and Schuster, 1984, p. 235-236
32 George Greenstein, *The Symbiotic Universe*, p. 38
33 *Grolier Multimedia Encyclopedia*, 1995
34 *Grolier Multimedia Encyclopedia*, 1995
35 The resonance mentioned here occurs as follows: when two atom nuclei fuse, the new emerging nucleus both takes on the total of the massive energy of the two nuclei forming it and their kinetic energy. This new nucleus works to reach a particular energy level within the atom's natural energy ladder. However, this is only possible if the total energy it receives corresponds to this level of energy. If it fails to correspond, then the new nucleus decomposes at once. For the new nucleus to attain stability, the accumulated energy in its body and the level of natural energy it forms should be equal to each other. When this equality is attained the "resonance" occurs. However this resonance is a highly rare harmony with a very low probability to be achieved.
36 George Greenstein, *The Symbiotic Universe*, p. 43-44
37 Paul Davies. *The Final Three Minutes*, New York: BasicBooks, 1994, p. 49-50 (Quoted from Hoyle)
38 Fred Hoyle, "The Universe: Past and Present Reflections", *Engineering and Science*, November 1981, pp. 8-12
39 Fred Hoyle, *Religion and the Scientists*, London: SCM, 1959; M. A. Corey, *The Natural History of Creation*, Maryland: University Press of America, 1995, p. 341
40 George Greenstein, *The Symbiotic Universe*,

p. 100
41 George Greenstein, *The Symbiotic Universe*, p. 100
42 George Greenstein, *The Symbiotic Universe*, p. 64-65
43 W. Press, "A Place for Teleology?", *Nature*, vol. 320, 1986, p. 315
44 Guy Murchie, *The Seven Mysteries of Life*, Boston: The Houghton Mifflin Company, 1978, p. 598
45 Michael Denton, *Nature's Destiny*, p. 11
46 George Greenstein, *The Symbiotic Universe*, p. 21
47 Jeremy Rifkin, *Entropy: A New World View*, New York, Viking Press, 1980, p. 6
48 Max Planck, May 1937 address, quoted in A. Barth, *The Creation* (1968), p. 144.
49 Paul Davies, *The Accidental Universe*, (1982) Cambridge: Cambridge University Press. Preface
50 Albert Einstein, *Letters to Maurice Solovine*, 1956, p. 114-115
51 Michael A. Corey, *God and the New Cosmology: The Anthropic Design Argument*, Maryland: Rowman & Littlefield Publishers, Inc., 1993, p. 259
52 G. W. Wetherill, "How Special is Jupiter?", *Nature*, vol. 373, 1995, p. 470
53 Michael Denton, *Nature's Destiny*, p. 262
54 F. Press, R. Siever, *Earth*, New York: W. H. Freeman, 1986, p. 2
55 See. Harun Yahya, *The Evolution Deceit: The Scientific Collapse of Darwinism and Its Ideological Background*, İstanbul, 1998.
56 Michael Denton, *Nature's Destiny*, p 106
57 F. Press, R. Siever, *Earth*, New York: W. H. Freeman, 1986, p 4
58 F. Press, R. Siever, *Earth*, New York: W. H. Freeman, 1986, p 4
59 F. Press, R. Siever, *Earth*, New York: W. H. Freeman, 1986, p 4
60 Michael Denton, *Nature's Destiny*, p.121
61 James J. Lovelock, *Gaia*, Oxford: Oxford University Press, 1987, p.71
62 Michael Denton, *Nature's Destiny*, p 127
63 Michael Denton, *Nature's Destiny*, p 128
64 Hugh Ross, *The Fingerprint of God: Recent Scientific Discoveries Reval the Unmistakable Identity of the Creator*, Oranga, California, Promise Publishing, 1991, p 129-132
65 Ian M. Campbell, *Energy and the Atmosphere*, London: Wiley, 1977, p.1-2
66 Ian M. Campbell, *Energy and the Atmosphere*, p.1-2
67 George Wald, "Life and Light", *Scientific American*, 1959, vol. 201, p.92-108
68 The near infrared range occupies the rays which extends from 0.70 micron, where visible light ends, to 1.50 micron.
69 This narrow range occupies the ultraviolet rays between 0.29 micron and 0.32 micron.
70 George Greenstein, *The Symbiotic Universe*, p 96
71 George Greenstein, *The Symbiotic Universe*, p.96-7
72 This chain reaction taking place in the eye is actually much more complicated. The light reaching the eye passes through the lens and falls upon the retina in the back. When light first strikes the retina a photon interacts with a molecule called 11-*cis*-retinal. The change in the shape of the retinal molecule forces a change in the shape of the protein, rhodopsin, to which the retinal is tightly bound. The protein's metamorphosis alters its behaviour. Now called metarhodopsin II, the protein sticks to another protein, called transducin. Before bumping into metarhodopsin II, transducin had tightly bound a small molecule called GDP. But when transducin interacts with metarhodopsin II, the GDP falls off, and a molecule called GTP binds to transducin.

Now, two proteins and one chemical molecule are bound to one another and it is called GTP-transducin-metarhodopsinII. It now binds to a protein called phosphodiesterase. When attached to metarhodopsin II and its entourage, the phosphodiesterase acquires the chemical ability to "cut" a molecule called cGMP. Initially there are a lot of cGMP molecules in the cell, but the phosphodiesterase lowers its concentration, just as a pulled plug lowers the water level in a bathtub.

Another protein that binds cGMP is called an ion channel. It acts as a gateway that regulates the number of sodium ions in the cell. Normally the ion channel allows sodium ions to flow into the cell, while a separate protein actively pumps them out again. The dual action of the ion channel and pump keeps the level of sodium ions in the cell within a narrow range.

When the amount of cGMP is reduced because of cleavage by the phosphodiesterase, the ion channel closes, causing the cellular concentration of positively charged sodium ions to be reduced. This causes an imbalance of charge across the cell membrane that, finally, causes a current to be transmitted down the optic nerve to the brain. The result, when interpreted by the brain, is vision. (Quoted from Michael Behe, Darwin's Black Box, New York: Free Press, 1996, pp. 18-21).

This is actually a very brief and simplified version of how we see. If the events developed like this, we would never be able to see. If the reactions mentioned above were the only ones that operated in the cell, the supply of 11-*cis*-retinal, cGMP, and sodium ions would quickly be depleted. There are many mechanisms that would restore the cells to their original state.

The reactions described above is far from being a complete biochemical explanation of seeing and they are only summarized. However, even what has been related above suggests that seeing is a very complicated and perfect mechanism which can never come about by evolution.

73 Michael Denton, *Nature's Destiny*, p 62, 69

74 Michael Denton, *Nature's Destiny*, p 55

75 *Encyclopaedia Britannica*, 1994, 15th ed., volume 18, p. 203

76 John Ray, *The Wisdom of God Manifested in the Word of Creation*, 1701; Michael Denton, Nature's Destiny, p. 73

77 Lawrence Henderson, *The Fitness of the Environment*, Boston: Beacon Press, 1958, Foreword.

78 The latent heat is the heat which does not change the heat of water but enables it to change it from solid state to liquid state or from liquid state to gas state. When you give heat to ice to melt it, the ice reaches to $0^{\circ}C$ and no increase in heat occurs even if you continue to heat it. Yet, it is no longer ice; it dissolves and becomes water. This heat, which is needed to convert the solid state into the liquid state despite causing no difference in temperature is "latent" heat.

79 Lawrence Henderson, *The Fitness of the Environment*, Boston: Beacon Press, 1958, p. 105

80 Michael Denton, *Nature's Destiny*, p. 32

81 Harold J. Morowitz, *Cosmic Joy and Local Pain*, New York: Scribner, 1987, p. 152-153

82 Michael Denton, *Nature's Destiny*, p. 33

83 Michael Denton, *Nature's Destiny*, p. 35-36

84 "Science Finds God", *Newsweek*, 27 July 1998

85 Robert E. D. Clark, *The Universe: Plan or Accident?*, London, Paternoster Press, 1961, p. 98

86 Fred Hoyle, *Religion and the Scientists*, London: SCM, 1959; M. A. Corey, *The Natural History of Creation*, Maryland: University Press of America, 1995, p. 341

87 David Burnie, *Life, Eyewitess Science*, London: Dorling Kindersley, 1996, p. 8

88 Nevil V. Sidgwick, *The Chemical Elements and Their Compounds*, vol 1. Oxford: Oxford University Press, 1950, p. 490

89 Nevil V. Sidgwick, *The Chemical Elements and Their Compounds*, vol 1., p. 490

90 J. B. S. Haldane, "The Origin of Life", *New Biology*, 1954, vol. 16, p. 12

91 Michael Denton, *Nature's Destiny*, p. 115-116

92 Lawrence Henderson, *The Fitness of the Environment*, Boston: Beacon Press, 1958, p. 247-48

93 L. L. Ingraham, "Enzymic Activation of Oxygen", *Comprehensive Biochemistry*, (ed. M. Florkin, E. H. Stotz), Amsterdam: Elsevier, vol. 14, p. 424

94 The question of how the complicated enzyme system enabling oxygen intake by the respiratory system emerged is one of the questions the theory of evolution fails to explain. This system has an irreducible complexity, in other words, the system can not function unless all of its components function perfectly. For this reason, it is unlikely to say that the system developed from the simple form to the more complex, as evolution suggests. Prof. Ali Demirsoy, a biologist from Ankara Hacettepe University and a prominent advocate of the theory of evolution in Turkey, makes the following confession about this subject:

"However, there is a major problem here. Mitochondria use a fixed number of enzymes during the process of breaking (with oxygen). The absence of only one of these enzymes stops the functioning of the whole system. Besides, energy gain with oxygen does not seem to be a system which can proceed step by step. Only the complete system performs its function. That is why, instead of the step by step development to which we have adhered so far as a principle, we feel the urge to embrace the suggestion that, all the enzymes (Krebs enzyme) needed to perform the reactions of the mi-

tochondria entered a cell all at once by coincidence or, were formed in that cell all at once. That is merely because those systems failing to use oxygen fully, in other words, those systems remaining in the intermediate level would disappear as soon as they react with oxygen." (Ali Demirsoy, The Basic Laws of Life: General Zoology, Volume 1, Section 1, Ankara, 1998, p.578)
While the probability of the formation of only one of the enzymes (special proteins) Prof. Demirsoy mentions above, saying "we have to accept that they formed all of a sudden by coincidence" is 1 over 10^{950}, it is certainly unreasonable to put forward that many enzymes of that sort formed by coincidence

95 Nevil V. Sidgwick, *The Chemical Elements and Their Compounds*, vol 1. Oxford: Oxford University Press, 1950, p. 490
96 Michael Denton, *Nature's Destiny*, p. 122-123
97 Irwin Fridovich, "Oxygen Radicals, Hydrogen Peroxide, and Oxygen Toxicity", *Free Radicals in Biology*, (ed. W. A. Pryor), New York: Academic Press, 1976, p. 239-240
98 J. J. R. Fraústo da Silva, R. J. P. Williams, *The Biological Chemistry of the Elements*, Oxford: Oxford University Press, p. 3-4
99 J. J. R. Fraústo da Silva, R. J. P. Williams, *The Biological Chemistry of the Elements*, p. 5
100 Michael Denton, *Nature's Destiny*, p. 79-85
101 Jeremy Rifkin, *Algeny*, New York: The Viking Press, 1983, p. 114
102 Robert Shapiro, *Origins: A Sceptics Guide to the Creation of Life on Earth*, New York, Summit Books, 1986. p.127
103 Fred Hoyle, Chandra Wickramasinghe, *Evolution from Space*, New York, Simon & Schuster, 1984, p. 148
104 Fred Hoyle, Chandra Wickramasinghe, *Evolution from Space*, p. 130
105 Michael Behe, *Darwin's Black Box: The Biochemical Challenge to Evolution*, New York, The Free Press, 1996, p. 252-53
106 Colin Patterson, "Evolution and Creationism", Speech at the American Museum of Natural History, New York (November 5,1981); Henry Morris, *That Their Words May Be Used Against Them*, AR: Master Books, 1997, s. 128
107 Hugh Ross, *The Fingerprint of God*, p.50
108. Charles Darwin, *The Origin of Species: By Means of Natural Selection or the Preservation of Favoured Races in the Struggle for Life*, London: Senate Press, 1995, p. 134.
109. Derek A. Ager. "The Nature of the Fossil Record." *Proceedings of the British Geological Association*, vol. 87, no. 2, (1976), p. 133.
110. T.N. George, "Fossils in Evolutionary Perspective", *Science Progress*, vol.48, (January 1960), p.1-3
111. Richard Monestarsky, Mysteries of the Orient, *Discover*, April 1993, p.40.
112. Stefan Bengston, Nature 345:765 (1990).
113. Earnest A. Hooton, *Up From The Ape*, New York: McMillan, 1931, p.332.
114. Stephen Jay Gould, Smith Woodward's Folly, *New Scientist*, 5 April, 1979, p. 44.
115. Charles E. Oxnard, The Place of Australopithecines in Human Evolution: Grounds for Doubt, *Nature*, No. 258, p. 389.
116. Richard Leakey, *The Making of Mankind*, London: Sphere Books, 1981, p. 116
117. Eric Trinkaus, Hard Times Among the Neanderthals, *Natural History*, No. 87, December 1978, p. 10, R.L. Holoway, "The Neanderthal Brain: What was Primitive?", *American Journal of Physical Anthropology Supplement*, No. 12, 1991, p. 94
118. Ali Demirsoy, *Kalitim ve Evrim* (Inheritance and Evolution), Ankara: Meteksan Yayinlari 1984, p. 61
119. Ali Demirsoy, *Kalitim ve Evrim* (Inheritance and Evolution), Ankara: Meteksan Yayinlari 1984, p. 61
120. *Fabbri Britannica Science Encyclopaedia*, Vol. 2, No. 22, p. 519
121. Kevin McKean, *Bilim ve Teknik*, No. 189, p. 7
122. Frank B. Salisbury, "Doubts about the Modern Synthetic Theory of Evolution", *American Biology Teacher*, September 1971, p. 336.
123. Ali Demirsoy, *Kalitim ve Evrim* (Inheritance and Evolution), Ankara: Meteksan Publishing Co., 1984, p. 39.
124. Homer Jacobson, "Information, Reproduction and the Origin of Life", *American Scientist*, January, 1955, p.121.
125. Reinhard Junker & Siegfried Scherer, "Entstehung Gesiche Der Lebewesen", *Weyel*, 1986, p. 89.
126. Michael J. Behe, *Darwin's Black Box*, New York: Free Press, 1996, pp. 232-233.

Al-Attique Publishers. Inc.Canada
65- Treverton Drive Scarborough , ON . M1k 3S5
Tel: (416) 615-1222 F: (416) 615-0375

Our Publications

No	Name of Book	Author`s Name	Page	C$	US$
122	The Bible led me to Islam ISBN: 9960-9148-3-6	Abdul Malik LeBlanc	82	3.00	2.00
123	Honor thy Father & Mother ISBN: 1-894264-17-7	----------------------	100	4.50	3.00
124	Why Islam is our only choice ISBN: 9960-9148-4-4	M. Hanif Shahid	291	12.00	8.00
125	Is Jesus God ? The Bible says No ISBN: 9960-9148-8-7	Shabir Allay	98	4.50	3.00
126	101 Questions to ask visiting Jehowah`s witnesses ISBN: 9960-9148-5-2	----------------------	93	4.50	3.00
127	101 Clear Contradictions in the Bible ISBN: 9960-9148-9-5	----------------------	39	3.00	2.00
128	Yahweh, Jehowah or Allah – Which is God's real Name? ISBN: 1-894264-18-5	----------------------	40	3.00	2.00
129	Source of Islamic theories ISBN: 9960-777-05-7	----------------------	20	1.50	1.00
130	What God said about Eating Pork. ISBN: 9960-777-02-2	----------------------	38	3.00	2.00
131	Common Questions People ask about Islam ISBN: 9960-9148-6-0	----------------------	75	4.50	3.00
132	Science in the Quran ISBN: 9960-9148-7-9	----------------------	42	3.00	2.00
133	Decision of the court : Quadianies are not Muslim ISBN: 9960-9036-3-x	----------------------	200	9.00	6.00
134	Follow Jesus of Follow Paul ISBN: 9960-777-06-5	Dr, Roshan Inaam	75	6.00	4.00
136	A Guide for Hajj and Umrah ISBN: 9960-9036-4-8	Anis & Daud Matthews	365	12.00	8.00

No	Name of Book	Author`s Name	Page	C$	US$
103	Hajj and Umrah (Urdu) ISBN: 9960-9148-1-x	Sh: Abdul Aziz Bin Baz	156	3.00	2.00
	Islamic Guidelines	M.Jamil Zino	190	8.25	5.50
	Pillars of Islam and Iman	---------------------	264	9.75	6.50
	Islamic Creed	---------------------	64	4.00	3.00
107	Credo Islamico (IslamicCreed) ISBN: 1-894264-02-9	----------------------	70	6.00	4.00
	What a Muslim Believes	M.Jamil Zino	68	3.00	2.00
109	Actions are by intentions ISBN: 1-894264-03-7	Immam :Sh Faisal	48	4.00	3.00
118	Ramadaan Companion- Spirtual Reflections ISBN: 1-894264-29-0	Sheikh Faisal Abdur-Razak	172	9.00	6.00
110	Help yourself in Reading Quran + 2 audio Cassettes ISBN: 1-894264-01-0	-----------------------	165	18.00	12.00
111	The Book of Death ISBN: 1-894264-05-3	-----------------------	162	7.50	5.00
112	The Universe seen through the Quran ISBN: 1-894264-00-2	Dr..Mir Anees-ud-Din Ph. D	167	10.50	7.00
113	Pearls of the truth 99-Name of Allah ISBN: 1-894264-2-8	Sh: Sidheeque M.A	208	10.50	7.00
114	Doomsday::Portents & Prophecies ISBN: 1-894264-06-1	-----------------------	494	27.00	18.00
115	Forty Hadith	Immam Nawawi	132	4.50	3.00
116	Monotheism and Pantheism ISBN: 1-894264-27-4	Amir Hamza	96	6.00	4.00
149	The Relationship between Muslim & Non-Muslim ISBN: 1-894264-30-4	Dr .Saeed Ismael Sini	125	6.00	4.50
150	Fate – Al-Qadaa Wal-Qadar ISBN: 1-894264-31-2	-------------------------	96	7.00	5.00
99	(Lesfilles De I`autre Voie) Daughters of an other Path (French) ISBN: 1-894264-28-2	Carol L.Anway Translated by: Renee Rivard	285	17.95	13.95

No	Name of Book	Author`s Name	Page	C$	US$
154	Daughters of Another Path : Experiences of American woman choosing Islam ISBN: 1-894264-0-9	Carol L. Anway	224	17.95	13.95
210	West Versus Islam ISBN: 1-894264-20-7	Margaret Marcus	98	6.00	4.00
153	The Fundamentals of ISLAMIC CREED ISBN: 1-894264-32-0	Abu Jaifer-Al-Tahewi	255	20.00	15.00
146	I would like to Repent But----- ISBN: 1-894264-37-1	M.Saleh Al-Munajid	53	3.00	2.00
148	Islam The Religion you can no longer Ignore ISBN: 1-894264-34-7	Syed Iqbal Zaheer	53	3.00	2.00
152	Bilal : The Abyssinian Outrunner ISBN: 1-894264-36-3	-------------------------	104	7.00	5.00
151	Muhammad : The Unlettered Prophet who changed the world in 23 years ISBN: 1-894264-33-9	-------------------------	102	6.00	4.50
147	A Voice to Hear ISBN: 1-894264-35-5	-------------------------	117	3.00	2.00
	نيبنيي نج يم تجتح ه				
41	يمنمنميتحيبه يمنمنمي تجيلم /تجبنفي نيبنيلمتج برنج تميمتح يم يفي		116	3.00	2.00
42	يمتمنمبم تحبننمه ينجينمنم برتم بر فيتحمم برنج يمنم		156	5.00	4.00
43	—تحتحتح بريميي برتميبنتم تحيمي برييمتحنم		272	12.00	9.00
44	تحتحتح فييمي يميتحيميفي		239	11.00	8.00
45	تحتحتح نمنييمبه برتحييتح تج تحيمي نيبه برنيينمتمي		264	12.00	9.00
46	تحتحتح ن... نيبنيمني تجيمبميمنم يمنمنمه		184	11.00	8.00
47	نج تج بن نمينمله برنج تمي تحيمي يبمنم برتم نمبنتحنج بريني بريمنمي نملجمم تحيم		103	5.00	4.00
48	نييبنبمنينج برنج نجتمتح		95	4.00	3.00
	نيبنيي نج بريتحلم				
49	ييميمنمي تجيلم نيي تجنمبمنم نمينمينمي تجنبميم نمي يي نج		128	3.00	2.00
50	بريميمنينمي نيبن تجنمي بربنبمنم نمينميبنيميمنينمي نييملميتحهنمي		62	2.00	1.00
51	برتحينجنميم برتحلملانمي يلميتمنمي لحيتحفيه تجبنفينمي نيبن تجنمي تمله / تحيم نج تج		148	9.00	6.00

May Allah (SWT) protect all of us from all evils.

Website: www.al-attique.com
E-mail: al-attique@al-attique.com & quran@istar.ca